THE BIOLOGY OF POLLUTED WATERS

THE BIOLOGY OF
Polluted Waters

by

H. B. N. HYNES

Professor of Biology,
University of Waterloo, Ontario, Canada

WITH AN INTRODUCTION BY

F. T. K. PENTELOW

Chief Inspector of Salmon and Freshwater Fisheries
Ministry of Agriculture, Fisheries, and Food
London, England

UNIVERSITY OF TORONTO PRESS

1974

First published 1960 by
LIVERPOOL UNIVERSITY PRESS

Reprinted 1960, 1963, 1966, 1971, 1974

Published in Canada and the United States by
UNIVERSITY OF TORONTO PRESS
Toronto and Buffalo

ISBN 0 8020 1690 1

Printed in Great Britain at the
University Printing House, Cambridge
(Brooke Crutchley, University Printer)

PREFACE

THIS book has been written in the hope that it will be of service to all those concerned with the problems of water pollution, but who are not specialists in freshwater biology. It is hoped that it will explain to such people as River Board officials, industrial managers and scientists, sewage works managers, civil engineers and fishermen the biological aspects of a problem which has become of nation-wide importance in Britain during recent decades, and which is also causing great concern in many other parts of the world.

It is not intended to be in any sense a text-book, its aim being merely to explain in fairly simple terms the extent of our present knowledge, and to indicate, by means of references, where more detailed information can be obtained.

The literature on the biology of polluted waters is wide and scattered, and in many languages, and much of it has been printed in publications which are not readily obtainable through ordinary library channels. The annual output of new papers is very large, as can be seen from the extremely useful *Water Pollution Abstracts* prepared by the Department of Scientific and Industrial Research and published by H.M. Stationery Office, or in the abstracts published in the American journal *Sewage and Industrial Wastes*. It is therefore impossible for one individual to read all that has been written on the subject, and I make no claim to having done so. I have, however, read many hundreds of papers from many parts of the world, and I hope that I have not overlooked any major works. The bibliography at the end of the book is a selection from my reading list, and the individual works have been included either because they contain observational data or because they deal with particular aspects or ideas which have been touched upon in the text. I hope that my fellow biologists may find that this attempt to survey and select from the literature will be of use to them.

It has been my experience, when discussing the fascinating subject of river biology with people other than biologists, that while many of them are quite familiar with the various kinds of plants and animals which are to be found in the water they have no names to apply to them. This makes mutual understanding difficult and the biologist's contribution to the discussion only partly intelligible. In order to overcome this difficulty I have included drawings of most of the invertebrates and microscopic plants which have been mentioned in the text, and also photographs of the common sewage-fungus organisms. I would emphasise here, however, that the drawings were made as pictures, not as guides to accurate identification of the creatures depicted. Specific differences, which from an ecological point of view are usually of great importance, are often small and of a kind which cannot be depicted in a drawing showing the general appearance of the organism. Precise identification of species is a task which, generally speaking, can be undertaken only by biologists, and even among them no one individual is capable of dealing with more than a limited number of groups of plants or animals.

It is a pleasure to acknowledge the great amount of help I have received from many people. This book owes its existence to the inspiration of Professor R. J. Pumphrey, F.R.S., who has given me unstinted help and encouragement throughout its preparation and helpful suggestions for its improvement. Dr. P. S. Dixon and Dr. R. O. Brinkhurst have read the manuscript and made useful comments. To these members of the University of Liverpool and to my wife, Mrs. M. E. Hynes, who has also been very helpful, I offer my grateful thanks. Many of my scientific colleagues have sent me copies of their works, and for this help I am particularly indebted to Dr. B. A. Southgate of the Water Pollution Research Laboratory and Dr. E. A. Thomas of the Kantonales Laboratorium, Zürich. Professor Kaj Berg of the Ferskvands Biologiske Laboratorium Hillerød, Denmark, has also helped me greatly by allowing me to read his *Furesøundersøgelser* while it was still in manuscript and the unpublished thesis of one of his past students, Mr. J. Birket-Smith. The latter work is a critical study of the German 'Saprobiensystem' and I found it a rich source of ideas and comments, many of which have been used in Chapter XIII.

It would not have been possible to undertake this work without the active assistance of librarians. Mrs. G. W. Flinn and Miss E. Whelan of the University of Liverpool, and Miss N. H. Johnson of the Water Pollution Research Laboratory, have been particularly helpful in obtaining papers and checking references. I am also very grateful to three members of the staff of the Department of Zoology, University of Liverpool: Mrs. A. Peat has done all the typing needed in the preparation of a book, Miss J. Venn undertook most of the tedious work of sorting and measuring specimens from my collections from the Afon Hirnant and has assisted me in many other ways, and Mr. W. Irvine took the excellent photographs which form the two plates.

I wish also to express my thanks to the Town Clerk of the Borough of Luton, Mr. A. D. Harvey, for permission to publish data which I collected from the River Lee at his request.

Finally I am very grateful to Mr. F. T. K. Pentelow for writing the introduction to this book. I am also indebted to him, and to Dr. T. T. Macan, Secretary of the International Association of Theoretical and Applied Limnology, for permission to reproduce *Fig.* 1. Acknowledgements for factual data which have been reproduced in figures and tables are given in the legends.

H. B. N. Hynes

Liverpool,
April 1959

CONTENTS

Preface		v
Introduction		xiii
Chapter I	THE HISTORY OF WATER POLLUTION	1
Chapter II	NATURAL WATERS AND NATURAL QUALITIES	9
Chapter III	ECOLOGICAL FACTORS AND RIVER FLORAS	18
Chapter IV	RIVER FAUNAS	27
Chapter V	EFFLUENTS AND CHEMISTRY	53
Chapter VI	PHYSICAL AND CHEMICAL EFFECTS OF EFFLUENTS ON RIVERS	64
Chapter VII	BIOLOGICAL EFFECTS OF POISONS	70
Chapter VIII	BIOLOGICAL EFFECTS OF SIMPLE DE-OXYGENATION AND SUSPENDED SOLIDS	86
Chapter IX	BIOLOGICAL EFFECTS OF ORGANIC MATTER	92
Chapter X	FURTHER BIOLOGICAL ASPECTS OF ORGANIC POLLUTION	122
Chapter XI	HEAT, SALTS AND POLLUTION OF LAKES	136
Chapter XII	OTHER HUMAN INFLUENCES ON NATURAL WATERS	146

Chapter XIII THE BIOLOGICAL ASSESSMENT OF POLLUTION 155

Chapter XIV THE PROSPECTS FOR THE FUTURE 168

References 176

Subject Index 191

Author Index 200

LIST OF ILLUSTRATIONS

Plates

1 Sewage Fungus *facing page* 96

A & B Two common growth forms of *Sphaerotilus natans* attached to pieces of grass

C *Fusarium aqueductum* growing on a dead twig

D *Apodya lactea* (*Leptomitus lacteus*) growing on a piece of moss

E *Beggiatoa alba* attached to a lump of mud

F *Asellus aquaticus* overgrown with a coat of *Zoogloea ramigera* and *Sphaerotilus natans*

G *Nemoura erratica* (a stonefly) nymph overgrown with a coat of *Sphaerotilus natans* in which many vegetable fibres are entangled

2 Microphotographs of sewage fungus and other micro-organisms *facing page* 97

A *Stigeoclonium tenue*—a green alga

B *Fusarium aqueductum*

C *Apodya lactea*

D *Carchesium polypinum*, as preserved in formalin

E *Zoogloea ramigera*

F The massive form of *Zoogloea* which occurs on sewage filter beds

G Filaments of *Sphaerotilus natans* mixed with *Zoogloea*

H Part of E under high power magnification

I Part of F under high power magnification

J *Sphaerotilus natans*

K *Beggiatoa alba*

L *Leptothrix ochracea*

FIGURES IN THE TEXT

Fig. 1 Folding Map showing the distribution of fishes in English and Welsh rivers. Redrawn from Pentelow 1955. 6

Fig. 2 Common algae from the plankton of various types of lake. 12

Fig. 3 The annual temperature cycle of the Afon Hirnant, Merionethshire, at different altitudes. 20

Fig. 4 Sessile algae which are important in the ecology of normal and polluted rivers. 25

Fig. 5 Flatworms, segmented worms, leeches, mites and mayfly nymphs which are important inhabitants of eroding substrata in running water. 29

Fig. 6 Stonefly nymphs, beetles and their larvae, and shrimps which are important inhabitants of eroding substrata in running water. 30

Fig. 7 Caddis-worms, the pupal case of *Agapetus*, and molluscs of eroding substrata in running water. 31

Fig. 8 Dipteran larvae, and the pupa of *Simulium*, of eroding substrata in running water. 33

Fig. 9 Diagram illustrating the life cycles of a typical stonefly and a typical mayfly. 34

Fig. 10 Invertebrates of depositing substrata in running water. 37

Fig. 11 Polyps, moss-animalcules, segmented worms, crustaceans, caddis-worms and molluscs of solid objects and weed beds of depositing substrata in running water. 39

Fig. 12 Mayfly and dragonfly nymphs, water-boatmen, beetles and fly larvae of weed beds on depositing substrata in rivers, lakes and ponds. 41

Fig. 13 The numbers of animals caught month by month during 1955–6 in the Afon Hirnant by a standardised netting technique. 49

Fig. 14 Histograms demonstrating the difference in percentage size-distribution of samples of *Baetis* and *Leuctra* collected with nets of coarse and fine mesh. 51

Fig. 15 Diagrammatic presentation of the decrease in concentration of a poison in a river and the corresponding changes in numbers of algae and numbers of species of animals. 82

Fig. 16 Diagrammatic presentation of the effects of an organic effluent on a river and the changes as one passes downstream from the outfall. 94

Fig. 17 The principal members of the 'sewage-fungus' community. 99

Fig. 18 Changes in the numbers of algae growing on glass slides in three organically polluted rivers. 102

Fig. 19 Changes in the composition of the principal members of the algal flora growing on glass slides in three organically polluted rivers. 104

Fig. 20 Protozoa which are important in the study of organically polluted water. 110

Fig. 21 Fly larvae which may occur in the septic zone of grossly organically polluted rivers. 112

Fig. 22 Diagrammatic presentation of the effect of a heated effluent on an organically polluted river. 138

INTRODUCTION

THE scientific study of polluted waters has been carried on in this country for about seventy years. It began when the West Riding Rivers Board was set up in 1894 and was intensively prosecuted by the Royal Commission on Sewage Disposal, which began work in 1901 and published its last report in 1915.

For many years primary emphasis was laid on chemical and bacteriological aspects, and two men, Frankland and Houston, became world famous for their studies in these fields. Apart, however, from bacteriology, which was of course the most urgent and pressing need in order to safeguard human health, the study of biology lagged behind chemical investigation. Nevertheless, the importance of biology soon made itself apparent, and biologists other than bacteriologists were employed both by the Rivers Board and by the Royal Commission.

These early workers were faced with a difficult problem because of the lack of information on the freshwater biology of this country, and it was not until a real interest in freshwater biology, which culminated in the foundation of the Freshwater Biological Association, awakened after the First World War that very much progress could be made. Developments since 1930 can be traced pretty clearly in Dr. Hynes's bibliography, and it is apparent that up and down the country a good deal of effort has been put into the study of the biology of polluted waters. Another obvious feature is the extent to which the observations are scattered over the country and the records of them through a variety of scientific journals. The seeker after information has therefore been faced with a formidable task in looking it up in the literature, and it has been extremely difficult to get any co-ordinated picture of the biological effects of pollution.

In this book Dr. Hynes has summarised and codified the information that is available. He has himself done a good deal of work on the problems he is discussing, and he is therefore well qualified by experience as well as by study to assemble the jigsaw to produce an intelligible and coherent picture. This is the first time this has been done in an English or American book.

It cannot be doubted that the biological effects of pollution are of great interest and importance to mankind. In a few cases it is true that pollution may be so severe and of such a nature that man or animals are poisoned by drinking the water, but those cases are few indeed as compared with those where the risk is of the transmission of water-borne diseases, of difficulties of providing potable water, and the destruction of fisheries; all biological phenomena.

Accordingly the study of polluted water becomes essentially ecological. What is the relation between the nature of the pollution and the nature of the community of organisms, animal, plant and bacterial, which develops in the water affected? How do the organisms composing these communities react on each other, and are their inter-relations affected by the abnormal physical and chemical factors in the environment? Then how do living organisms affect the environment of polluted waters? Some of these studies are old and some new, and they all find their place in this book.

To say that all these problems have been elucidated would be going a good deal too far, but by the critical examination of a wide range of facts assembled from all over the world a great deal of light is thrown on them. For example, many effluents from many different sources contain solid matter in suspension, sometimes with dissolved matter as well, and sometimes without; and a study of the modes of life of a variety of organisms shows how it is that some can tolerate and even use silt and others can not, and also goes some way to disentangling the effects on animals and plants of complex effluents like sewage, showing what is likely to be due to physical factors and what to chemical.

This book provides not only an indispensable summary of existing knowledge but also a firm base for new thought and new advances in our knowledge of the biology of polluted waters. I am very honoured that Dr. Hynes should have asked me to write an introduction to it.

F. T. K. PENTELOW

Chapter I

THE HISTORY OF WATER POLLUTION

POLLUTION is a word that occurs often in the newspapers and is seldom for long off the tongues of freshwater fishermen; of recent years it has also come into prominence in the law courts. According to Wisdom (1956), pollution is legally definable as 'the addition of something to water which changes its natural qualities so that the riparian owner does not get the natural water of the stream transmitted to him.' This has the appearance of being a clear and concise statement; but what is the natural water of a stream? And what are its natural qualities?

Ever since man ceased to be a roving hunter, and probably even before he took up settled life in communities, he has altered the landscape by clearing trees, burning and cultivating, and by piling up middens—all activities which lead inevitably to alterations in streams. The clearing of forests increases the rate of surface run-off and so results in more violent fluctuations in the speed of flow of rivers; such fluctuations are, as we shall see later, an important factor in river biology. Cultivation leads to soil erosion and so to more silt in the streams; and the middens of Stone Age man, like those of his present descendants, must have increased the amount of putrescent organic matter reaching the water. So water pollution, even if one defines it in purely human terms, is undoubtedly older than history.

Even 'natural' streams may show the characteristic signs of pollution. In densely wooded regions the autumn leaf-fall may add so much organic matter to water that fish are asphyxiated. Schneller (1955) has investigated this effect in an American stream, and the reader will be familiar with the foetid appearance of many woodland streams and pools in this country. In such places the water is murky and smells foul when disturbed, and the decaying leaves near the surface are covered with a white coat of sewage fungus. These conditions occurring near a cesspool or a town-dump would immediately be attributed to gross pollution by human agency. There is little doubt, however, that they have occurred in some places in every age since the invasion of the land surface by plants, and they must have been widespread in the coal-forests of the Carboniferous Period.

'Natural' pollution may also faithfully reproduce the effect of the addition of industrial poisons to water. Huet (1951) has shown that small trout-streams in the Belgian Ardennes are adversely affected by afforestation with spruce and red cedar. The invertebrate fauna is reduced and the trout disappear from the little head-streams in which they normally spawn; the fishing is consequently impaired in the larger rivers many miles below the plantations. This effect is due in part to loss of food, for the shade limits the growth of algae and the tough coniferous needles are a poor substitute for the soft leaves of trees like alder, which, when shed into the water, form the food of many aquatic invertebrates. But there is also some toxic effect, for the damage persists below the plantations and it is more marked in streams which arise from marshy areas than in those which rise as springs. Some noxious substance, as yet unidentified, dissolves out of the layer of needles on the marshes and seeps into the water. It is fortunate that circumstances made it possible for Huet to exclude direct human agency, and to attribute the degeneration of the fishery to its real cause. Had any mine or factory existed in the area it is probable that its effluent would have been the first suspect, and the investigation might not have been pursued further.

There is no doubt, however, that at the present time the most serious pollution is the direct result of human activity. As soon as large settlements and towns became common, the problem of the disposal of domestic waste arose. Earth-closets and the burial of excrement have been sanitary measures from Biblical times, but many ancient cities had elaborate systems of sewers. These last must have resulted in severe pollution if we are to judge by nineteenth-century experience in Britain. It may be noted that Aristotle mentions the white colour produced by foul mud and the small red threads that grow out of it; sewage fungus and sludge worms were therefore known to the ancients (Liebmann, 1951).

With the decline of the Roman Empire much of the world reverted to the older methods and allowed the soil to undertake the purification of buried waste. In Britain this probably served well enough in the Middle Ages, and even as late as the seventeenth century the population remained small and scattered. Izaak Walton, whose *Compleat Angler* was first published in 1653 and of which the fifth edition appeared in 1676, made no mention of pollution. Indeed, it is from his account that we know that some of our rivers, which are now in very poor condition, were then full of fish. If pollution had been recognised then as a serious problem he could hardly have written a book on freshwater fishing without mentioning it—unless indeed the angling fraternity has changed its nature. Nowadays anglers are very conscious of, and very vocal about, pollution. Had Walton, however, visited some parts of Cornwall, the centre of early mining operations, he might well have had something to say on the subject. We know from Herodotus that tin was mined in Britain in the fifth century before Christ, and Lovett (1957) has discovered what is probably the earliest reference to pollution in England: Spenser's *Faerie Queene*, written in the sixteenth century, describes the Dart as 'nigh choked with sands of tinny mines'. Where there is tin there is usually lead and zinc, and we know from modern works that 'sands' from such mines produce devastating effects.

During the eighteenth century the state of the towns was disgusting; garbage and excrement accumulated in the streets, and as the population increased, so did the filth. At the same time new industries arose, and more and more contamination found its way into the rivers. Early in the nineteenth century the introduction of sewers and water-carriage of excrement increased the load, and by the middle of the century, although the streets were cleaner, there were increasing complaints about the state of many rivers. People began writing letters to *The Times* on the subject and have continued to do so at intervals ever since.

At that time the sewage was discharged into the rivers completely untreated. Fortunately this practice has been discontinued, but even now, over a century later, untreated, or only very partially treated, sewage is still discharged in large amounts into tidal waters. There are, however, signs that public feeling is being aroused about this, as it was about the rivers. In 1957 several seaside resorts became aware of the dangers involved and posted notices warning people of the risk of infection with poliomyelitis if they bathed in the sea. The realisation that marine pollution is not only unpleasant but that it may also endanger public health may well have beneficial effects. Recently the South Wales Sea Fisheries District Committee has urged that full treatment should be given to sewage discharged into tidal waters, and the city of Bristol is planning full treatment of the sewage it discharges into the estuary of the Avon.

It was early in the nineteenth century that it became clear that the condition of the rivers was getting out of hand. There were outbreaks of cholera in London, and salmon disappeared from the Thames and the Mersey. Obviously something had to be done, and the result was a series of Acts of Parliament of varying efficacy. The Gas Works Clauses Act of 1847 prohibited the discharge of gas-waste into streams, and the Salmon Fisheries Acts of 1861 and 1865 made it an offence to pollute salmon-waters so as to kill fish. As, however, the last salmon had already gone from many rivers these were only holding measures. A Royal Commission on Prevention of River Pollution was set up in 1857; it reported eight years later on methods of sewage treatment, and recommended the continuous application of sewage to land. An echo of their recommendation is often heard today when people talk of 'sewage farms', although this method of treatment has been abandoned almost everywhere. In 1865 another Royal Commission was established. It sat for many years and issued a series of reports on the state of various rivers, and it attempted to define standards, in physical and chemical terms, to which effluents should conform. The resulting Rivers Pollution Acts of 1876 and 1890 placed certain restrictions on the discharge of polluting matter into streams, other than tidal waters, but they had many defects, not the least of which were that they were cumbersome to administer and did not define polluting substances. Other Acts followed dealing with particular watersheds, areas or industries, but in the main it can safely be said that they did little to halt the degradation of the rivers. Finally, in 1898, the Royal Commission on Sewage Disposal was set up, and with its ten reports published during the first fifteen years of this century, the modern attitude to water pollution may be said to have begun. Unfortunately, probably because of the First World War, no Act of Parliament resulted, but the Commission's recommendations have been very widely accepted. They made excellent contributions in their scholarly discussions of the methods of treating effluents from sewers and some industries, but more important was their attempt to find some practical method of measuring pollution. They investigated a large number of rivers, and showed that the dissolved oxygen absorption test (now known as the five-day Biochemical Oxygen Demand or B.O.D.) gave a very fair measure of their cleanliness. This test, which is an arbitrary measure of the amount of putrescible material in the water, is discussed further in Chapter V. The Commission found that rivers could be classified as follows:

	B.O.D. (parts per million* dissolved oxygen absorbed in five days)
Very clean	1
Clean	2
Fairly clean	3
Doubtful	5
Bad	10

* The original figures were given in parts per 100,000, but nowadays it is customary to quote such figures in parts per million, which is the same thing as milligrams per litre (mg./l.).

They therefore recommended that a B.O.D. of 4 parts per million (p.p.m.) should not be exceeded and, assuming that a B.O.D. of 2 p.p.m. would occur in the average diluting water, they suggested that effluents which would receive an eight times dilution should not have a B.O.D. of more than 20 p.p.m. In the same way they suggested that such effluents should not contain suspended solids in excess of 30 p.p.m. Here for the first time the idea of a dilution factor was introduced, although this is often forgotten when the Royal Commission 20/30 standard is discussed. This standard has often been quoted in the law courts in cases brought under Common Law, but it has only had statutory recognition in a few Local Authority Acts (Wisdom, 1956).

In 1923 the Salmon and Freshwater Fisheries Act extended the protection given to salmon waters to those containing trout or coarse fish, and it gave Fishery Boards some control over the effluents which could be discharged into waters under their jurisdiction. This resulted in considerable improvement in some rivers and checked further deterioration in others, but the Act did not apply to rivers from which fish had already disappeared. It also left untouched the rather unsatisfactory state of affairs in which too many different authorities were responsible for each watershed; drainage and fisheries, for instance, remained under the control of different boards. The multiplicity of authorities had been criticised by the 1898 Royal Commission and was again attacked by the Joint Advisory Committee on River Pollution to the Ministers of Health and Agriculture and Fisheries. This body, which was set up at about the time of the passing of the 1923 Act, issued four reports between 1928 and 1937. Although it was primarily concerned with the administration of the existing law, it also made recommendations about the method of dealing with trade wastes. These led to the passing, in 1937, of the Drainage of Trade Premises Act, which gave industry the right, under certain conditions, to discharge effluents into sewers. This Act has greatly benefited our rivers, and will doubtless continue to do so, because many industrial effluents, which are difficult to treat by themselves, are much easier to purify when they are mixed with sewage (Collman, 1957). The Minister of Health next instituted the Central Advisory Water Committee, which again recommended, in 1943, the setting up of authorities responsible for administering all aspects of the management of entire watersheds. The resulting River Boards Act of 1948 established the thirty-two River Boards which, together with the much older Thames and Lee Conservancies, now administer the rivers of England and Wales. Similar bodies were set up for Scotland by the Prevention of Pollution (Scotland) Act of 1951.

The Central Advisory Water Committee was convened again in 1945, and it appointed a sub-committee to study the prevention of river pollution. This body reported in 1949 (Ministry of Health, 1949), and from its report arose the River (Prevention of Pollution) Act of 1951 and the similar Act covering Scotland, which has already been mentioned. Under these Acts the whole problem of administering the rivers and *of prescribing standards for effluents that can be allowed to enter them* is placed in the hands of the River Boards. This arrangement allows considerable flexibility; for instance, standards can be suited to local conditions, and it would seem that, given time, these bodies should be able to improve all our rivers. Further details of the law and its development are given in the excellent accounts of Pentelow (1953) and Wisdom (1956).

I think that 'given time' are words to which considerable importance should be attached: too many people expect miracles. One occasionally reads letters in the daily papers which seem to imply that the Authorities ought to be able to arrange that pollution can be turned off as one turns off a tap. Even learned judges seem to be under this impression, but the brief history of legislation given above shows that this simply cannot be done. The 1847 Act prohibited the discharge of gas-wastes into streams, but had it been immediately enforced the gas industry would have been killed outright. This is a densely populated and very industrialised country, and one can run neither industry nor sanitation without producing effluents, which must flow somewhere. It is often said that they should be piped straight to the sea, but that is usually both impracticable and undesirable. So much water is now used by man that in many places all or nearly all the water flowing in the lower reaches of rivers has been used for some purpose or other. There are so many wells and bore-holes in southern England that springs have dried up, and there are some rivers whose major source is now a sewage works. If the sewage were piped away, these rivers would dry up altogether except in wet weather. The truth is that, thanks largely to the work of the Water Pollution Research Laboratory of the Department of Scientific and Industrial Research, methods for the treatment of most types of effluent are available. These are described in detail by Southgate (1948), and go a long way towards providing solutions to the problems of pollution. But very few effluents can be rendered completely innocuous; most of them become so only if they receive reasonable dilution after treatment, and sewage is one of these. Treatment plants are expensive and until recently the materials necessary for their construction have been in short supply. Because of apathy, neglect, ignorance and rapacity, which have always been with us, there is an enormous leeway to make up. The two World Wars have added to the problem as they prevented any

FIG. 1. Map showing the distribution of fishes in English and Welsh rivers. Redrawn from Pentelow 1955.

attempt to redress the situation 'for the duration', whereas population and industries continued to expand.

Turing (1947–9) showed that in the years immediately after the Second World War the state of many rivers in Britain was very bad; and many of them are little better today. It is, however, easy to exaggerate the gravity of the situation, and this is done only too often by certain sectional interests. Doubtless some rivers will have to remain as main drains for a long time to come. New industries and new towns will present further problems, which will have to be tackled as they arise, and there will before long be the problem of radioactive wastes. Many rivers have, however, improved considerably in recent years as Local Authorities and industries have been able to catch up with their accumulated arrears in the installation of treatment plant. Pentelow (1955) strikes a hopeful note when he points out that most English and Welsh rivers still contain fish despite 200 years of industrialisation in a country where 43 million people live on 37 million acres. It is estimated that this population produces 1,500 million gallons of sewage per day (Southgate, 1957a), of which rather more than three-quarters, together with a similar proportion of all industrial wastes, are discharged into inland waters. The map from Pentelow's paper is reproduced here as *Fig.* 1 and it does much to restore confidence in the future of the rivers. He rightly stresses that no real judgement should be passed on anti-pollution efforts until they have been at work for about a century. We have just entered an age when town-planners and borough engineers really worry about pollution, and industrialists consider the disposal of effluents in deciding where to site new factories. Symptomatic of this changed attitude is the fact that when, recently, the Department of Civil Engineering of King's College, Newcastle, organised a two weeks' course on the prevention of river pollution it was attended by about one hundred people, mostly from industry (Isaac, 1957). Perhaps, after all, the angry protests of fishermen and land owners are bearing fruit, though it may not ripen in their lifetimes. But the brightest hope for the salvation of our rivers is the tranfer of responsibility, and a good deal of authority, to the River Roards. They alone can see each river as a whole and take into account such things as the quality of the diluting water, the degree of permissible impoundment and a host of other factors of which most other people are ignorant, or with which they are not concerned. It must be admitted that the record of riparian owners as defenders of their rivers is not irreproachable. Many have drawn large, sometimes very large, incomes from the lease of rights which they have not spent a penny to preserve. And when they discovered that they owned a wasting asset their first and only idea of a remedy has been an injunction and damages. Such feckless and irresponsible conduct cannot be permitted indefinitely to impair the spirit of co-operation which is necessary for the solution of what is, after all, a national problem. The River Boards can act more gently, but more effectively; they are making great strides in the control of pollution, and they have already achieved much without becoming involved in disputes which have had to be referred to higher authority (Pentelow, 1958).

There is one aspect, however, which, in my view, some of them have not yet sufficiently appreciated, and that is that pollution is fundamentally a biological problem. Almost every Board has a chemist on its staff, but few as yet employ biologists. I believe that the purpose for which these bodies were designed would be better and more speedily served if each employed a water chemist and a freshwater biologist who together could tackle the many problems involved in the most effective manner. Control of pollution requires the help of both disciplines working together in the closest co-operation.

Chapter II

NATURAL WATERS AND NATURAL QUALITIES

BEFORE considering the effect which pollution may have on thequalities of natural waters, it is necessary to attempt to define just what we mean by natural waters. It must be admitted at the outset that this is extremely difficult. In Chapter I it was pointed out that 'natural' waters may have been altered considerably by 'natural' pollution, but, in addition to this, many of the factors with regard to which normal waters differ from one another are just those on which various types of pollution exert the most influence.

Adequate definition and classification here would almost necessitate a treatise on the extensive science of freshwater biology. However, more than half that science is concerned with lakes, and in this country, and indeed in most others, lakes are not often polluted. This is partly because they tend to lie far from centres of civilisation and partly because they are clearly less suitable repositories for effluents than are rivers, which carry the offending matter away.

In two respects, however, pollution does affect lakes, and as both are concerned with productivity we must briefly review this aspect of lake biology. Large lakes may serve one or both of two biological purposes as far as man is concerned. They may support fisheries, as in Central Europe, where much of the table fish eaten locally is obtained from them, or they may serve as water supplies. In Britain the latter function is now by far the most important; but not so very long ago, before the railways and steam trawlers made sea fish available everywhere, lake fisheries had considerable local importance. There were, for example, large fisheries for the so-called 'lake herrings' in Loch Lomond, Lough Neagh and Llyn Tegid, and the potted char of Windermere were famous.

Lakes are essentially transitory things which ultimately fill up and disappear. In geological terms they do this very rapidly; and as they age they change (Welch, 1935). Very few are more than a few tens of thousands of years old, and those that are, such as Lake Baikal in Siberia and Lake Tanganyika, have survived only because of their vast size. In time even they will vanish, because every particle of silt which is brought into them settles to the bottom, and every creature which grows in them and dies helps to fill up their basins. In shallow water round the edges this process is accelerated by the growth of rooted aquatic plants which are commonly, but perhaps incorrectly, referred to as 'weeds'. These accumulate silt and cause the steady encroachment of the shores. At the same time there are chemical and biological changes which affect the open water.

Lakes are formed by some geological accident, such as subsidence, large-scale faulting, damming of valleys by rock falls or the gouging action of glaciers, and the initial properties of each lake depend on local conditions. If the drainage basin is rocky and infertile, very small amounts of the nutrient salts needed for plant growth, namely potassium, nitrates and phosphates, are carried into the lake: its waters are therefore infertile and its biological productivity low. If, on the other hand, the drainage basin is fertile and composed of easily weathered rocks and rich soils, relatively large quantities of nutrients find their way into the water, and the lake is productive. These two examples are the extremes of a continuously variable series, and limnologists use the terms oligotrophic and eutrophic for unproductive and productive lakes respectively. But an oligotrophic lake does not remain indefinitely in the same condition: not only does its basin change, as erosion occurs and more fertile soils are built up, but it also acts as a trap for fertility. This is shown, for example, by a table, reproduced below, from a booklet on the biology of water supply (Pearsall *et al.*, 1946), giving the average amount of saline nitrogen in the water entering and leaving four Lake District lakes. The concentrations are expressed as parts per million.

	Inflow water	Outflow water
Thirlmere	0·51	0·53
Windermere	0·71	0·55
Esthwaite Water	1·16	0·74
Lowes Water	0·62	0·35

It was estimated that Windermere gained about eight tons of nitrogen in this way during the year, and this, when built up into plankton, represents a *dry weight* of about 100 tons of organic matter. Such nutrients as do reach a lake tend, therefore, to become fixed in the bodies of plants and animals, and when these die and decay they form a sub-aquatic soil which slowly releases the nutrients for other lake-dwellers. So, in time, the length of which depends on local circumstances, the fertility becomes noticeably greater. The lake tends, in fact, to become more eutrophic. Similarly, a lake which was initially fairly fertile becomes richer with the passage of time. This has important effects on plants and animals; the total quantity of living matter increases, and different species succeed one another. These changes are of considerable scientific interest, but we are concerned with only two of them here.

Firstly the change from oligotrophy to eutrophy involves a great increase in the amount of plankton, the small plants and animals which live in the open water, and these affect the usefulness of the water to man. Oligotrophic lakes have a sparse plant-plankton, composed very largely of desmids, and correspondingly few planktonic animals. As fertility increases desmids are replaced by diatoms, and then by various flagellates and other green algae; finally, blue-green algae appear, and these can become so abundant at some seasons as to form 'water blooms'. Such outbreaks of plant-plankton growth can give the water the appearance of pea soup, a phenomenon which has been long known in Cheshire as 'the breaking of the meres'. The increase in tiny plants is followed by an increase in the small animals, mostly crustaceans, of the plankton. The resulting large amounts of algal and animal matter produce great problems for the water engineer and other general losses of amenity, which are discussed in Chapter XI. *Fig.* 2 shows some of the more common species of planktonic algae which occur in the waters of lakes, including some which cause trouble in water works even when they are not particularly abundant. Plankton can also affect the use of water for other purposes, for example industrial cooling. Considerable enrichment of lake water, which is a sympton of ageing, is therefore definitely a disadvantageous process from the human point of view.

A second change brought about by the ageing of lakes affects fisheries and, in those countries where lake fisheries are important, it can have serious consequences. Fish species, like those of the algae and invertebrates, change as lake waters become more fertile. Unfortunately the fishes which are most favoured by the gourmets are those of oligotrophic waters—such species as the trout, char and the various types of lake herring. At first these become commoner as the fertility of the water increases, but they are then slowly replaced by fishes like the perch, pike and minnow. As the process continues other species appear, such as the roach and bream, which are even less favoured for the table. In Switzerland these changes are causing grave concern, as in some lakes famous table fish such as the 'omble chevalier' (a large char) are beginning to disappear, and perch are taking their place. Those who have enjoyed the excellent dish the Swiss prepare from perch—boneless fillets, cooked

DIATOMS

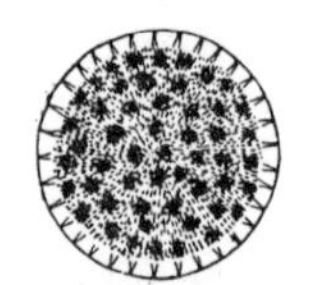

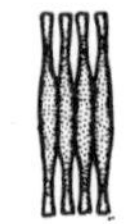

BLUE-GREENS

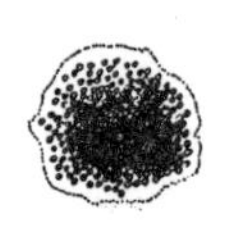

DESMIDS

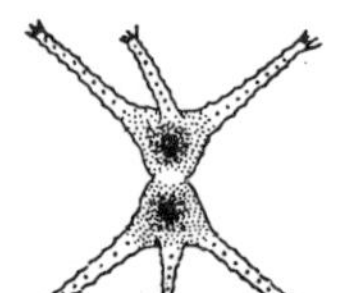

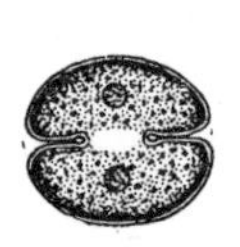

FLAGELLATES

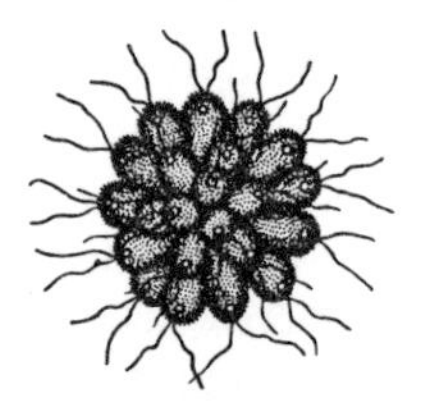

OTHER GREEN ALGAE

FIG. 2. Common algae from the plankton of various types of lake. Drawn to different scales as shown.

fast in deep fat so that they curl up and become almost crisp, served with chips—may wonder why anyone should worry, but it is doubtful if even Swiss cuisine could render palatable the disgustingly muddy flavour of roach, whose numbers may be expected to increase later as the process goes on.

As will be shown later, pollution, even by highly purified effluents, increases the rate of ageing of lakes (Pearsall, 1949); but it is questionable whether an increase in the rate at which a perfectly normal process occurs is an alteration in the natural quality of the water. The condition of an oligotrophic lake which is rendered more eutrophic by pollution is no more 'unnatural' than that of another lake which has reached the same state without human assistance. This point is rendered the more pertinent by the fact that, although they lie in general on a graded series of development, lakes present difficult problems of classification from a biological point of view. Indeed limnologists spend much time and energy arguing about their detailed classification. The general opinion is that, while the broad outlines are clear, each lake has its own individuality. Unless, therefore, alterations produced by pollution are very striking, *or detailed records of the previous condition and rate of change of a particular lake are available*, it is very difficult to assess how much of the alteration is natural and how much has been caused by human activities.

Turning now to rivers and streams, which are the more usual victims of pollution and so of paramount importance to us here, we have a much worse problem of classification. Indeed almost every biologist who has tried to make a rational classification has failed to produce one that works for any aspect other than the one which he happened to be studying. It may, however, at this stage be helpful to review some of the attempts, as they demonstrate the factors which control the occurrence of animals and plants in this type of environment.

Geographers have an ideal river. It rises in the mountains, and flows swiftly, over rocks and boulders, to the foothills, where it is still swift enough to carry sand and small stones along and so retain a gravel bed: when it reaches the plains, it tends to deposit silt and indulge in meanders. But many rivers, indeed most, do not conform in detail to this scheme, even if they happen to rise in mountains and flow through foothills on to plains (Macan and Worthington, 1951). Odd regions of hard rocks, sudden drops and other geological irregularities, cause swift and slow-flowing reaches to alternate, and one may find shallow gravelly riffles even where there are meanders. Many creatures of rivers and streams are controlled more by the type of bottom on which they live than by the general physical state of the river; thus, as a result of local variation, classification rapidly breaks down.

On the Continent it is common practice to divide rivers into zones which are characterised by certain fishes. A recent exposition of this system is given by Huet (1954), who lists four such zones:

1. *The trout zone*, with rapidly flowing well-oxygenated and cool water, and a steep-sided valley.

2. *The grayling zone*, with less rapid water and some slow reaches, although these are always well oxygenated, and with a steep-sided but flat-bottomed valley.

3. *The barbel zone*, with moderate current and long slow reaches: the oxygen content may fall in warm weather and a variety of fish species may be present. The valley is wide and flat-bottomed.

4. *The bream zone*, where the water is all slow-flowing and canal-like; the oxygen content may fall to low levels, and hardy fish like bream and tench occur.

This system has its merits, but it is difficult to apply in detail, and in Britain, where both the grayling and the barbel have restricted distributions resulting from geographical history, the system can hardly apply at all. Indeed even on the Continent it is applicable only with difficulty to anything other than fishes. For example Schmitz (1957) shows that while it is possible to find some correlation between the fish zones and the invertebrates, he is able to do so only by elaborate manipulation of the numbers of species occurring in each zone, and even then the results are neither very clear nor very convincing. Similarly Illies (1955) has shown that the upper zones of rivers in the Alps and Scandinavia are only vaguely comparable, because different species have different geographical distributions. He also points out that invertebrates tend to blur the boundaries between zones as no two species which occur together in one river have exactly the same distribution. Similarly any one species tends to behave differently in different rivers. While, therefore, it is possible to use such a zonation on a broad basis, it really tells one very little more about the detailed biology of the river than can be learned from a good map.

Carpenter's (1928) modification of the system for British conditions, in which she divides rivers into four reaches—the headstream, the trout-beck, the minnow reach (= approximately the grayling zone) and the lowland reach—suffers from the same objections, which she fully appreciated.

Butcher (1933) outlined a classification of river reaches on the basis of the larger plants that are to be found in them. This system is perhaps more useful than one based on fishes, because the plants are there to be seen and they do not move around and mislead one. Fishes are an unsatisfactory basis for classification, as they often move into areas where they really cannot maintain themselves; they are then no more inhabitants of the place where they occur than is the fly on the window-pane. Many readers will recall the reports in the newspapers of the death of large numbers of trout in small upland rivers and streams during the long drought of 1955. After such a disaster the stream fills again, and the plants and invertebrates are found still to be there, or at any rate most of them (Hynes, 1958a), but the trout may not return for months. Butcher's units of classification, of which there are five, are better considered as communities; this is because he does not relate them to their vertical position on the river system, but rather to the local conditions. He considers the most important factors to be current speed, the degree of silting, and to a lesser extent the hardness of the water.

Ricker (1937) also introduced the idea of considering local factors, in his classification of streams in Ontario, which he bases on local geology and soil, current speed, type of bed, temperature, volume of flow, flora and fauna, and oxygen and carbon dioxide content of the water. But even here, although there is more precision in that more factors are considered, one is still left with the impression that the classification works nicely on paper, and perhaps for the small area on which it was worked out. The broad general principles apply, but none of these schemes works in detail or very widely. For instance Ricker's first division, into creeks and rivers, is based on the volume of flow (more or less than 10 cubic feet per second) on June 1st. Any Briton knows that here June 1st may be the last day of a long May drought or there may be a raging flood, presaging the English summer holiday period.

One is very soon brought to the conclusion that rivers are strict individualists, each of which varies in its own way so as to make nonsense of anything but a very broad general classification. Such a classification can be made fairly readily at the level of the geography text-book, but as soon as one attempts a detailed study of the animals and plants it is the very local conditions, the sum of which ecologists call the microhabitat, that override all other considerations. In whatever fish zone it happens to occur, a patch of clean-swept gravel supports the sort of plants and animals peculiar to that type of microhabitat, and these creatures differ from those in similar places elsewhere in the river system only if there are other differences, such as temperature in hot weather or the hardness of the water. Indeed, Percival and Whitehead (1929) came to the conclusion that the type of substratum was the overruling controlling factor, a point to which I shall return. Similarly Berg (1948), at the end of his exhaustive studies of a Danish river, rejects all attempts that have been made to classify water courses as such, and concludes that the factors which control river animals are current speed, the type of substratum, the type of vegetation (which is, of course, like the last, a function of other factors), the temperature, the amount of oxygen in solution, the hardness of the water and finally the geographical position of the river. This last is the same point as that raised by Illies. All creatures have a limited geographical range, so one cannot expect always to find the same set of species in the same type of microhabitat. We, in Britain, for example, lack many species which occur on the Continent. They may have lived here once, but if so the Pleistocene Ice Age drove them out, and the formation of the English Channel has prevented their return. Others have managed to return, but some of them have not yet spread far. The grayling is a good example of this last group. It requires a moderate current and well-aerated but not very cold water, hence its normal habitat in the 'grayling zone'. During the Ice Age such British rivers as remained free of ice must have become too cold for it and it presumably died out. Then after the recession of the ice it returned northwards with the warmer climate, but being a fish, and a freshwater one which could not withstand sea water, it could only get to places to which it could swim. It moved down the Rhine from the south and arrived in the area of the North Sea, before the sea had risen to its present level, and when the Dogger Bank was still dry land. At this time many of the eastern rivers of England were confluent with the Rhine, and so the grayling could swim into them without entering salt water: but that is as far as it went. The rising sea level cut the species off from the Rhine and it remained confined to England's eastern rivers until quite recently, when man carried it around the country. Grayling now occur in some Welsh rivers, but this is due to human interference. Many other creatures have similarly restricted distributions, and this applies particularly to those freshwater animals which cannot move across land. This fascinating subject has recently been discussed in great detail by Thienemann (1950).

Returning now to the classification of water courses it is clear that no system which has been devised, or is likely to be, can be expected to apply in detail. It is possible to work out a general system for fishes, if one takes geographical history into account, and possibly also one for rooted plants, but when it is necessary to consider the whole biological community, as in the study of the effects of pollution, one must deal with microhabitats, and even then geographical history is important. Moreover many of the important qualities in which natural river reaches differ from one another, such as type of substratum, amount of silt, oxygen content of the water, hardness of the water and temperature, are just those which as we shall see, are altered by most types of pollution. It is evident therefore that pollution may merely change one type of 'natural' water into another, and that this applies to rivers as well as to lakes. Only when pollution is so severe as to cause changes which overstep the boundaries of normal variation is it possible to say that the water has lost its 'natural qualities'.

Chapter III

ECOLOGICAL FACTORS AND RIVER FLORAS

It was concluded in the last chapter that classification of river environments can only be satisfactorily made at the level of the microhabitat. The largest workable unit is the small, fairly uniform, reach, and such reaches are normally arranged in a rather haphazard manner depending on local geology. There are a number of environment properties which in various ways make up the ecological factors of the local microhabitat. These have indeed been the basis of the general attempts at classification of rivers which have already been discussed and are of basic importance to the biological communities of rivers. They are:

1. *The dissolved salts in the water*, the most important of which is calcium bicarbonate leached out of lime-bearing rocks. These are not only limestones and chalk but many other types of rock such as basalt and marls. Where the water flows from non-calcareous rocks, such as slates, shales or granite, very little calcium is present in solution; the water is therefore soft and may be acid. Acidity occurs particularly where the landscape is peaty; because the water is quite unbuffered the acids added from the peat are not neutralised: in physico-chemical terms the pH is lowered. Hard waters on the other hand are alkaline, and their pH does not change readily; if acid is added, say from the moors which often develop on areas of drift on limestone hills, it is neutralised by the calcium bicarbonate in solution. Suppose for instance one adds some hydrochloric acid to hard water, the following reaction occurs:

$$Ca(HCO_3)_2 + 2HCl \rightarrow CaCl_2 + 2H_2O + 2CO_2.$$

The hydrochloric acid is replaced by an equivalent amount of carbon dioxide, which is a very weak acid, and the pH is little changed. Thus hard water, quite apart from the fact that it contains calcium salts which are needed by some animals for shells or other skeletal structures, is more stable chemically than is soft water.

Other dissolved salts, or more strictly speaking ions, which are important are those of potassium, nitrate and phosphate which are needed for plant growth; traces of these are always present even in rain water (Gorham, 1958), but they are also leached out of the rocks and soil, and the more fertile the drainage basin the more ions will be present in the water. On the whole the softer lowland rocks and soils produce more than the mountain-forming rocks.

2. *The speed of the current* is important, not only directly, but indirectly, as it influences the type of river bed and the amount of silt deposition. Several authors, among them Macan and Worthington (1951), have quoted a table showing the relation between current and the nature of the river bed (Table 1). This gives a general indication, but in prac-

TABLE 1

Relation of current speed and nature of river bed. Reproduced from Butcher (1933)

Velocity of current per second	Nature of bed	Type of habitat
More than 4 ft. (1·21 m.) . .	rock	torrential
More than 3 ft. (0·91 m.) . .	heavy shingle	torrential
More than 2 ft. (0·60 m.) . .	light shingle	non-silted
More than 1 ft. (0·30 m.) . .	gravel	partly silted
More than 8 in. (0·20 m.) . .	sand	partly silted
More than 5 in. (0·12 m.) . .	silt	silted
Less than 5 in.	mud	pond-like

tice, of course, the current is not uniform all over the river bed nor is it invariable. Also the presence of a few scattered larger stones amongst shingle causes it to be more stable, and so alters its character as a biological habitat.

3. *The temperature of the water* is clearly dependent on altitude, aspect and the type of source (deep spring or surface run-off). It is, in practice, only the higher temperatures which are critical. All rivers are cold in winter, but in Britain none of them freeze solid. In summer, however, sluggish lowland rivers may become very warm, occasionally reaching temperatures as high as 25° C., whilst high streams remain cool. A series of temperature readings made at six points on a hill stream in Wales (*Fig.* 3) illustrates this point. It can be seen that, even within the confines of a small trout-stream, altitude has an important influence on the temperature régime. This stream is a tributary of the Dee, and if the readings had also been extended to the lowland reaches of that river the annual temperature range would have been found to be much greater still. These graphs also show that the winter temperatures were usually higher at the higher altitudes. The explanation is that the temperature of the springs, which in this case contribute most of the head-water,

varies little with the season and depends only on the mean annual temperature, whereas the surface run-off water, which is added in increasing proportion as the stream flows down, is the product of the local weather at the time. At all altitudes spring waters have a remarkably even temperature, and one often finds creatures in them which are intolerant of the high summer temperatures of nearby streams.

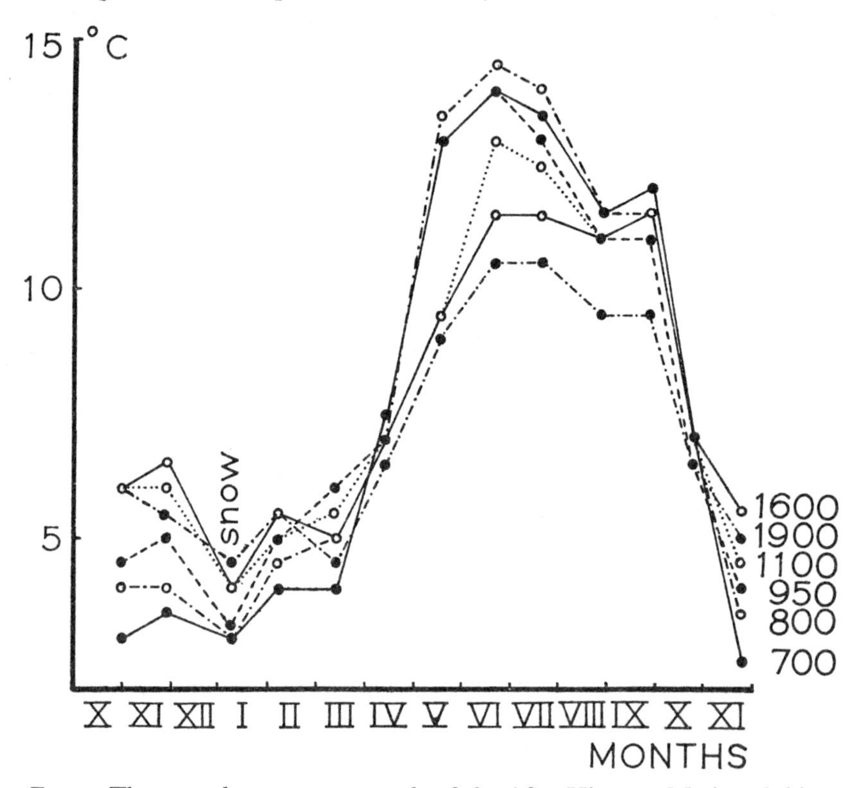

FIG. 3. The annual temperature cycle of the Afon Hirnant, Merionethshire, at different altitudes. The figures on the right are the altitudes of the six stations in feet. Based on monthly measurements made during 1955–6.

4. *The dissolved oxygen content of the water.* Many animals, notably trout and many invertebrates, can live only in well-aerated water, but it is not the actual amount of dissolved oxygen which matters so much as the percentage saturation. This point often leads to misunderstandings between biologists and chemists, as the latter are used to measuring things in terms of their concentration per unit volume. Oxygen is not very soluble in water, and its solubility depends on the temperature. At normal atmospheric pressure (760 mm. mercury), and taking into account

the fact that air contains only 20·9 per cent. of oxygen, the 100 per cent. saturation value varies from about 14 p.p.m. at 0° C. to zero at boiling point. The saturation values for water in contact with pure oxygen are, of course, nearly five times as great. These facts have important consequences in stream biology. Living things, and their decay, use up oxygen which can only be replaced from the atmosphere or by the activities of green plants. In a swift upland stream the movement and turbulence of the water rapidly replenishes any oxygen used and the water remains saturated. If it tends to become over-saturated, either because of a rise of temperature or because of photosynthesis by plants in bright sunshine, oxygen is lost to the atmosphere; thus at all times the percentage saturation remains around 100 per cent. In a sluggish river, however, there is less turbulence and oxygen is not so readily absorbed from, or given up to, the air. Thus the amount in solution may fall considerably, particularly at night. It is prevented, however, under ordinary circumstances, from falling to very low levels by the fact that the rate of uptake increases as the percentage saturation falls. This rate is proportional to the saturation deficit; if the rate of oxygen uptake under given conditions is x when the saturation deficit is 10 per cent. (i.e. the percentage saturation is 90 per cent.), it becomes $2x$ when the deficit is 20 per cent. (80 per cent. saturation). The relative inability of sluggish water to exchange oxygen with the atmosphere can also result in the phenomenon of supersaturation. Under conditions of bright sunshine the plants produce a large amount of oxygen, and this, being pure oxygen, can theoretically result in nearly five times the concentration which could be dissolved from the air at the same temperature. In practice 500 per cent. saturation is never attained, but values up to 200 per cent. have been recorded. In a turbulent stream, where the easy exchange with the atmosphere tends to keep the balance, the percentage saturation with oxygen is therefore always high; in a sluggish stream values tend to fluctuate widely, and the lowest values are reached at night or in dull weather (Butcher *et al.*, 1927). The warmer the water the greater the rate at which oxygen is used up and the less there is in solution to use, so that very low values tend to occur mainly on summer nights.

These then are the physical and chemical factors which determine the type of habitat which any river offers to living things. River and stream waters may be hard or soft, fertile or infertile, swift-flowing or sluggish, and warm or cold; all combinations and intermediates may occur, and any individual river may vary in respect of these characteristics along its length.

At the base of all living communities lie the plants, both large and small, and these have been especially investigated in this country by Dr.

Butcher. His study of the large plants (1933) will serve as a basis for our consideration of the living inhabitants of rivers. Unlike land plants, those that grow in rivers are never short of water but they may be short of light, for 60 per cent. of the incident light may be absorbed by 6 feet of apparently clear water, and much more if the water is turbid. They would appear also, at first sight, to be short of carbon dioxide, the main raw material of photosynthesis, as this is often less abundant in water than it is in air. However, except in very soft waters, there is usually an adequate amount of the bicarbonate ion HCO_3, which can be used by plants with equal facility. Aquatic angiosperms, or flowering plants, apparently absorb their nutrient salts from their roots and not directly from the water, so they, like land plants, are dependent on the sort of soil which is available to them. This does not apply to mosses, which have no roots in the ordinary sense.

Butcher has shown that in running water the current is of paramount importance, and that different communities of plants are found as one passes from swift- to slow-flowing water. The structure of the plants is to some extent correlated with the rate of flow of the water, those which grow in torrential habitats being either cushion-shaped or provided with strong flexible stems. After quoting the table shown here as Table 1, he defines five principal plant communities, which are listed below with some modifications.

1. *Torrential*, on rock or heavy shingle; river mosses, especially *Fontinalis antipyretica* and *Eurhynchium riparioides*. Other species of moss may also occur, e.g. *Hypnum* spp., and in my experience in the very soft waters of North Wales *Fontinalis squamosa* replaces *F. antipyretica*. Another important plant in this community is the alga *Lemanea*, which often forms large moss-like growths, as does the alga *Vaucheria* in fertile water.

2. *Non-silted* on shingle; water crowfoot (*Ranunculus fluitans*) and river milfoil (*Myriophyllum spicatum*) together with water celery (*Berula*) in harder water. Again in my experience *M. spicatum* is replaced in softer water by the smaller species *M. alterniflorum*, and *Callitriche* (starwort) is often found in this community.

3. *Partly silted* on gravel and sand; *Ranunculus* still present, but accompanied by various species of pond weeds (*Potamogeton*) and also in harder water, by simple burr reed (*Sparganium simplex*), mare's tail (*Hippuris*) and water soldier (*Sagittaria*).

4. *Silted* on silt; starwort (*Callitriche*), other species of *Potamogeton* and Canadian pond-weed (*Elodea canadensis*), and one should add here the yellow water lily (*Nuphar lutea*).

5. *Littoral* on mud where there is a very little current. This is a very

varied community. Butcher lists only the grass *Glyceria aquatica* and burr reed (*Sparganium erectum*), but many others may occur, such as the bull rush (*Schoenoplectus lacustris*), mace reed (*Typha* spp.) and the common reed (*Phragmites*). In soft waters this community may be composed of rushes (*Juncus* spp.) or sedges (*Carex* spp.), or even, in acid water, of bog moss (*Sphagnum* spp.), and in small streams where the water is fairly hard the banks may be lined by water cress (*Nasturtium officinale*) or water parsnip (*Apium nodosum*). This large community is probably best referred to as the 'emergent' rather than the littoral community. It is not always littoral, and may occur far from the shore in shallow water, but it is always recognisable as the plants stand up out of the water.

A notable feature about these plant communities is that they do not occur everywhere; there are nearly always bare areas due to scour, periodic drought or other factors, and the individual patches of plants expand and contract and move around. River weeds, like land plants, also alter their environment by forming soil. Their decay provides soil-building material and they trap silt and build up mud banks. These can be produced very rapidly; Butcher records that a large clump of *Ranunculus fluitans* raised the level of a river bed by 17 cm. (6½ in.) between May and September. Such piling up of silt may, of course, not be stable, and the whole edifice may be swept away by floods, carrying the original plant with it. This sort of impermanence is one of the reasons for the rapid and often spectacular changes shown by the plant communities of running water. Undoubtedly the general stability of the river bed and the amount of fluctuation in current play an important part in the life of river weeds. It can be seen in almost any upland stream that rolling stones gather no moss, which is confined to fixed boulders or solid rock surfaces. Jones (1943) has surmised that at least one of the reasons for the lack of plants in the river Rheidol in Cardiganshire is the shifting nature of the bed. In comparing this river with the similar and nearby river Teifi, which has a rich flora of *Ranunculus*, *Myriophyllum* and *Callitriche*, he points out that the bed of the Rheidol is covered with a constantly shifting layer of lead-mine spoil which must prevent the establishment of plants.

Large plants, or macrophytes, terms which in this context include angiosperms, mosses and large conspicuous algae, are not, however, the only plants of running waters, nor indeed are they the most important. The surfaces of stones, rocks and even of the macrophytes themselves are densely covered with millions of algae. Not only are these tiny plants a major source of food for animals, far more important in this respect than macrophytes (Percival and Whitehead, 1929), but they are probably more important than the macrophytes as oxygenators of the water. Butcher (1933) finds that they produce super-saturation more readily than large plants and suggests that this is a direct consequence of their small size, for they produce very tiny bubbles of oxygen which readily dissolve in the water, whereas large plants produce larger bubbles which rise and burst at the surface. The latter can be seen very easily if a portion of a plant such as *Elodea* is exposed to bright sunlight in a glass vessel.

The algae are difficult to study *in situ*. Some work has been done on scrapings from stones, and one worker (Margalef, 1949) has suggested the ingenious idea of covering areas of stone with celloidin which, when hardened, can be peeled off, taking the algae with it. But most work on the algae of rivers has been done by suspending glass microscope-slides in the water for a period and then examining them to see what has grown. This method is open to various objections; it is, for instance known that some algae never appear on such slides. It may be that a glass surface, even if ground, does not supply the right type of substratum, but it is more probable that it is a matter of time of immersion. Some algae, such as *Lemanea*, never seem to grow on the broken bottles which are unfortunately so common in our streams, but others do appear after a long time. It is, for instance, possible to find *Hildenbrandia*, a red encrusting plant, on bottles in Windermere. This alga does not often grow on slides, and the bottles on which it now appears are the old-fashioned 'ginger-pop' bottles with marbles in their necks. This type of bottle went out of use many years ago, so they must have been in the lake for a very long time. However, despite the objections to it, the glass-slide method has been widely used and has the advantage of being simple and of permitting a quantitative comparison of the productivity of different waters.

Butcher (1946a, 1948) has reported very valuable results from many rivers. He found that in oligotrophic (infertile) waters (note that this term can be used to refer to rivers as well as to lakes) the number of algae growing on the slides varied between about 1,000–5,000 per square millimetre (about half a million to three million per square inch!). In eutrophic waters the numbers rise to 10,000/sq. mm., and can on occasion reach 100,000, although such large numbers seem to be unstable and tend to fluctuate widely. In all rivers there are seasonal changes in abundance; there is a minimum of algae in winter and maxima in spring and autumn, the two maxima being caused by different species. There are also different species in oligotrophic and eutrophic waters. In the former, especially near the source, many species are characteristic, although they may be scarce, but one nearly always finds the diatoms *Eunotia* spp, *Achnanthes* spp. and *Diatoma haemale*, and often *Ceratoneis arcus*, *Tabellaria* spp. and members of the Chaetophorales. As one proceeds downstream the water becomes more eutrophic and the algal community changes until it becomes dominated by the diatom *Cocconeis placentula*, and the green algae *Chamaesiphon* spp. and *Ulvella frequens*, and several other species of diatom are usually present also. These include *Synedra ulna*, *Navicula viridula*, *Surirella ovata*, *Cymbella ventricosa* and *Gomphonema olivaceum*. Many of these algae are illustrated in *Fig*. 4. The change in dominants occurs whether the water is hard or soft, although there are differences in the subsidiary species, and the controlling factor seems to be the amount of available nutrient salts. It would seem that the *Cocconeis/Chamaesiphon/Ulvella* community represents an end point

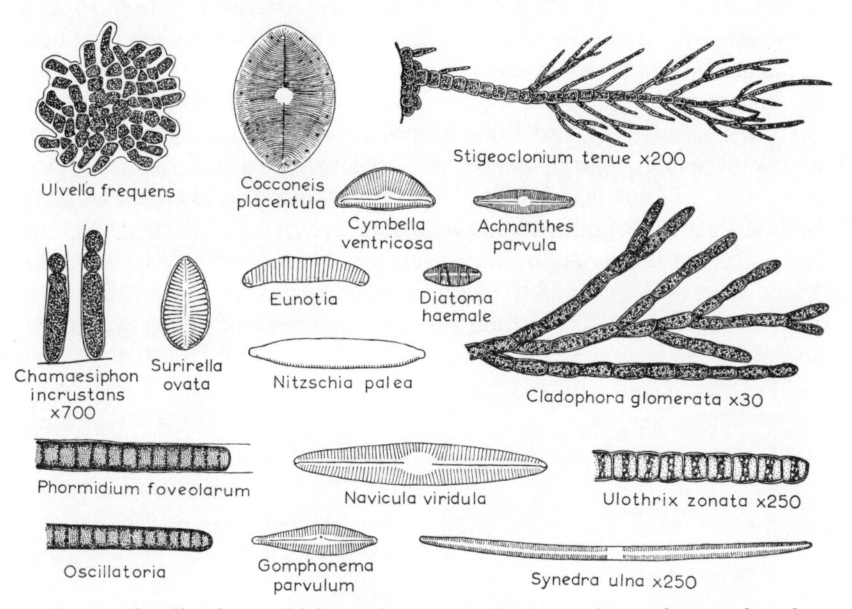

FIG. 4. Sessile algae which are important in the ecology of normal and polluted rivers. Magnified x 500 except where shown otherwise.

in the development of river algal communities; it is in fact a *climax* in the sense in which this term is used by the plant-ecologist, and it appears wherever there is a suitable substratum and an adequate supply of nutrient salts. It develops best—that is, it reaches maximum numbers —in places where the flow is swift, perhaps because it is intolerant of the smothering effect of silt.

The algal community of rivers is essentially sessile, it grows on solid bodies and can develop only where these are present; in places where the substratum is soft mud it can grow only on weeds or hard parts of the bank. In muddy reaches also the steady rain of silt on to the bottom smothers tiny plants; silt also greatly lowers the light intensity and renders algal growth less possible. In such reaches the algae which do occur, other than those living on weeds, live in the open water and are therefore planktonic, as are the algae of the open water of lakes. But river plankton is not indigenous, it cannot maintain itself in the way that lake plankton does, as, except in very large rivers like the Nile and the Danube, it is carried away so fast by the water that no individuals remain for long in the river.

If one samples the water of a head-stream with a fine-meshed net one finds a few algae in the open water, but these are all members of the bottom flora scoured off the stones by the current (Butcher, 1932b). Downstream, where the river may be draining lakes or ponds or where there are pond-like backwaters, some typical lake-plankton algae, such as are shown in *Fig.* 2, may occur. These have come into the river from their normal habitats, and although still alive and in some circumstances abundant, as for instance where there are large lakes on the course of the river, they are not native to it. The amount and kind of such plankton, both of algae and of planktonic animals, therefore depends very much on the sources of the water and varies very much from river to river. On the whole it is probably of little significance in river biology, although it may serve as a source of food for invertebrates and so influence the abundance of such species as can take advantage of it (Illies, 1956).

Chapter IV

RIVER FAUNAS

THE animal inhabitants of rivers are controlled not only by the four factors listed in Chapter III, but also by the plants, which provide them with shelter as well as with food. And, odd though it may seem, river invertebrates, and through them ultimately the fishes, are also influenced by terrestrial vegetation. This is because many of the small animals eat vegetable detritus, such as dead leaves and grass stems and the products of their breakdown and decay. In fact this terrestrial debris and the algae are the major basic foodstuffs of the animal community of rivers. Relatively few animals actually eat the tissues of the aquatic macrophytes, as is readily apparent, not only from the study of the gut-contents of the animals, but from the plants themselves. In summer-time almost every terrestrial plant, indeed almost every leaf, shows signs of damage by insects, snails or slugs, and the same applies to most plants taken from ponds. But examination of river weeds reveals very few signs of damage even when they are full of insects, shrimps or snails. Some caddis-worms make their cases of bits of leaf (e.g. *Halesus*, *Fig.* 11) and some midge-larvae burrow into the leaves for shelter (Walshe, 1951), but otherwise few creatures actually damage plant tissue. The main function of the plants in the provision of food is that they increase the surface available for the growth of attached algae, which are grazed by the animals.

The major controlling factor in the distribution of invertebrates is the nature of the river bed. This is either 'eroding' in which event it is rock, stones or gravel, or 'depositing', in which event it is silt or mud. The intermediate condition, sand, forms a convenient dividing line as it is a particularly unsuitable habitat for animal life, and is often almost barren. Superimposed on any of these types of bottom one may find weed beds, which provide a third type of substratum.

Both the oxygen content of the water, and its temperature, tend to be correlated with the type of river bed. Eroding substrata occur in turbulent water and usually in the upper reaches of rivers or near springs in the lowlands, so they are well oxygenated and usually cool. Depositing substrata, on the other hand, occur only in sluggish water and usually at low altitudes; they are therefore liable at times to be deficient in oxygen, and in summer-time to be warm. Extensive weed beds tend to occur in silted regions of slower current (see Chapter III) and are similarly occasionally exposed to oxygen deficiency and high temperatures. Beds of *Ranunculus*, *Myriophyllum*, *Callitriche* or mosses often occur, however, in swift cool reaches, so the fauna of weed beds to some extent cuts across our primary division.

The other important factor affecting animals, the hardness of the water, is not so closely correlated with the substratum, although even here there is some agreement. Swiftly flowing waters, with eroding substrata, may be either hard or soft, but the more sluggish lowland reaches of rivers, which have depositing substrata, have on the whole relatively hard water. This is because they flow through areas of softer rocks, which supply at least some calcium to the water. In practice the biological distinction between soft and hard water lies at a much lower level than that of the housewife, who thinks, or rather before the advent of detergents used to think, in terms of soap-suds. The dividing line between soft and hard water from the biological point of view is about 20 mg./l. of calcium. The housewife would consider this rather soft, and would not begin to complain seriously until this figure had become 60 or even 80 mg./l. Apparently few river animals are affected adversely by hard water, but some, particularly those which form shells, require at least a minimum amount of calcium, which seems to be around 20 mg./l., and they become more common the harder the water. Such creatures are the shrimp *Gammarus pulex* (*Fig.* 6) and most of the snails and bivalves. For example Jones (1948a) found that all these animals were absent from several soft-water streams in South Wales, although they were present in large numbers in the nearby hard-water stream the Clydach. There are also indications that some of the insects are similarly affected, although why this should be so is obscure. One can understand why a mollusc or a shrimp, which uses calcium for its shell, should need a minimum concentration, but even this is not a complete explanation. Some molluscs, most notably the river limpet *Ancylastrum fluviatile* (*Fig.* 7) and several species of pea-mussel, *Pisidium* (*Fig.* 10), occur in very soft waters, although they are much less abundant there than in similar hard waters, and *Gammarus pulex* has occasionally been found in waters containing less than 4 mg./l. of calcium. Clearly we have still much to learn about the calcium physiology of such creatures, and the figure of 20 mg./l. is an arbitrary one which is perhaps more correlated with abundance than with occurrence.

The inhabitants of the 'eroding' river beds are a rich assortment of worms, leeches, shrimps, insects, mites and molluscs, some of which are illustrated in *Figs.* 5, 6, 7 and 8. They have varied ways of life, but

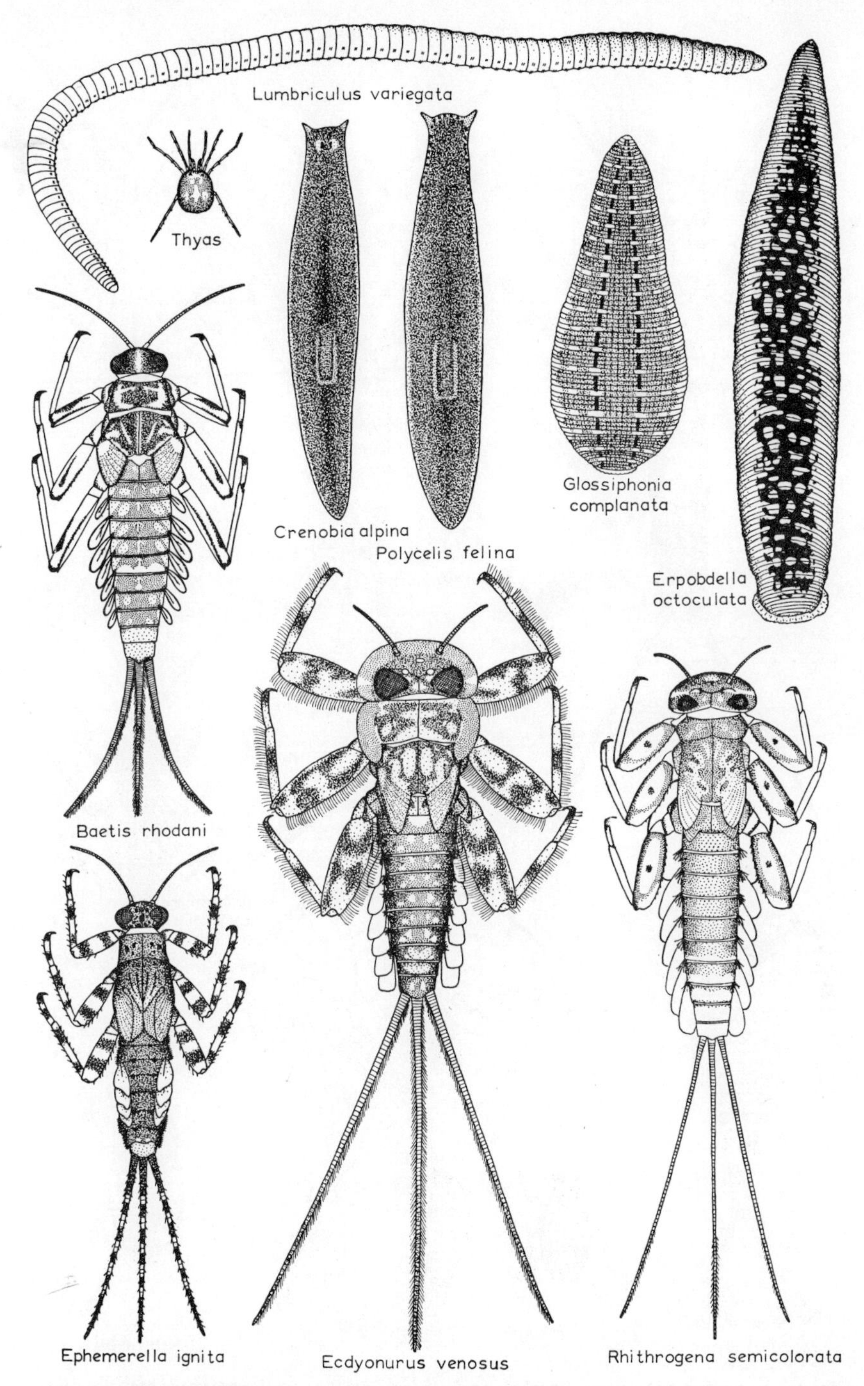

FIG. 5. Flatworms, segmented worms, leeches, mites and mayfly nymphs which are important inhabitants of eroding substrata in running water. Magnified x 5. *Ephemerella ignita* is also a common animal of weedbeds in turbulent water and many of the animals shown here also occur on stony lake shores.

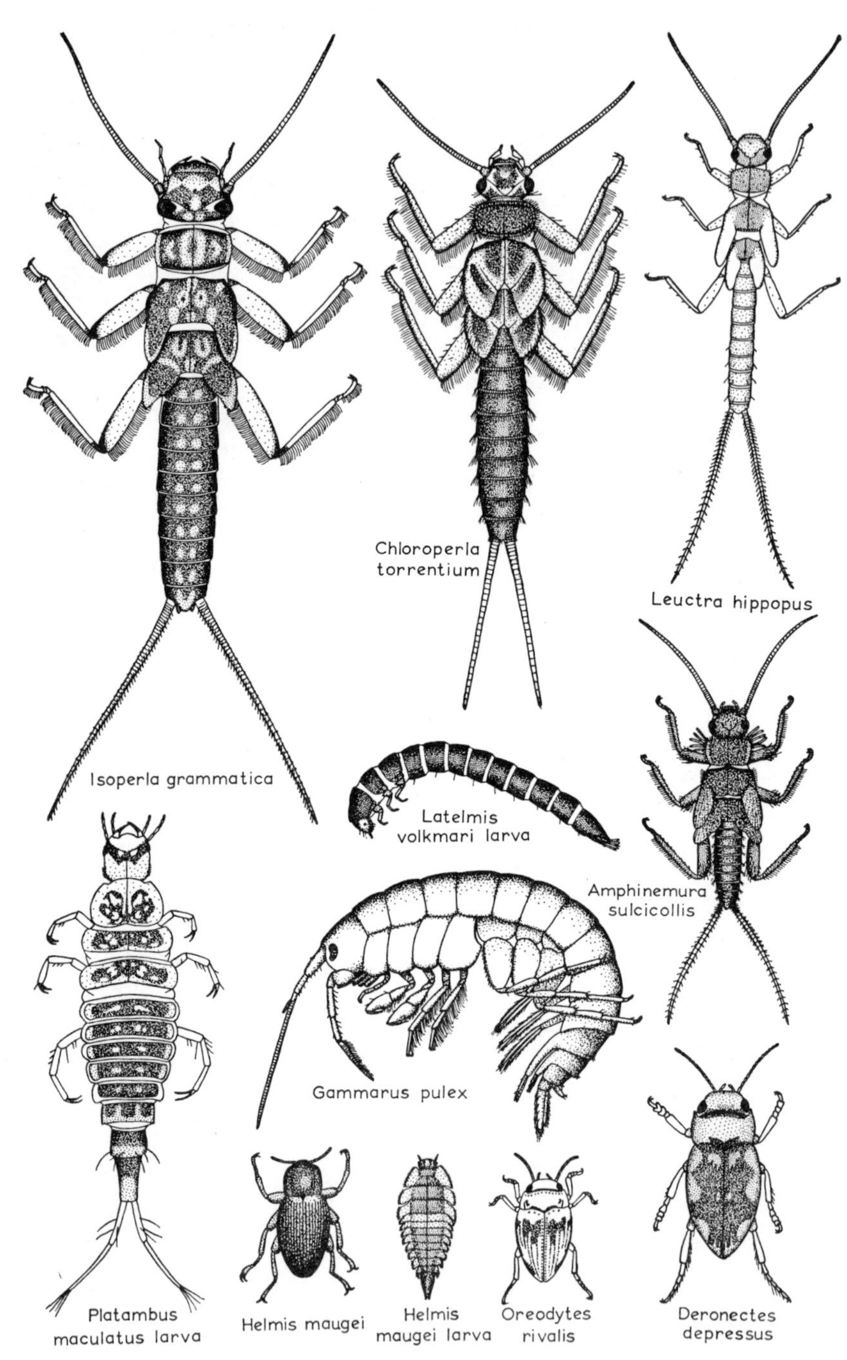

FIG. 6. Stonefly nymphs, beetles and their larvae, and shrimps which are important inhabitants of eroding substrata in running water. Magnified x 5. *Isoperla grammatica*, *Amphinemura sulcicollis*, *Helmis maugei* and *Gammarus pulex* become more abundant where weeds or moss provide cover.

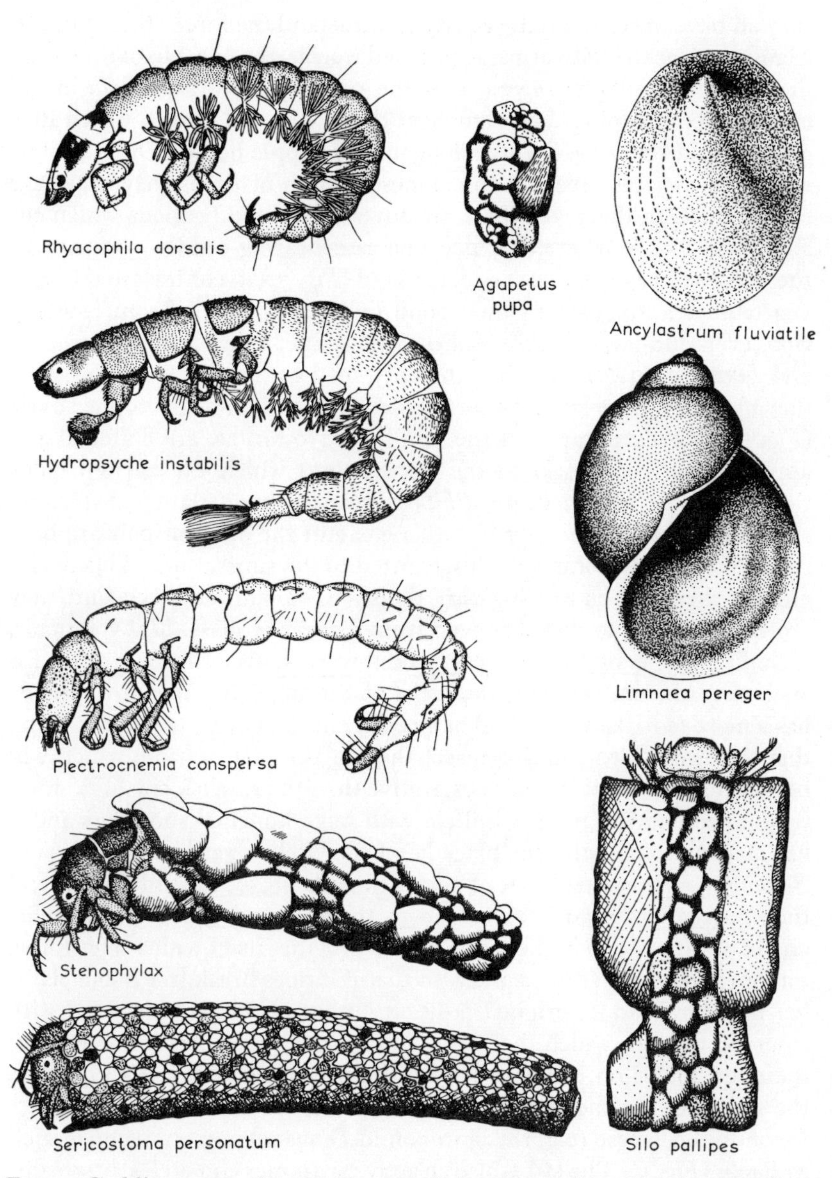

FIG. 7. Caddis-worms, the pupal case of *Agapetus*, and molluscs of eroding substrata in running water. Magnified x 5. *Limnaea pereger*, of which only the shell is shown, is found also in many other types of habitat.

they all have in common the ability to withstand the force of the current. Many, such as the flatworms, segmented worms, the nymphs of the stoneflies *Leuctra* and *Chloroperla* and the cranefly larva *Dicranota*, avoid exposure by creeping down amongst the stones and gravel, which they are able to do because of their elongate and supple bodies. Others habitually live exposed lives on the stones, though at times they may take shelter beneath them. All these creatures show modifications which enable them to live where they do. The case-bearing caddis-worms make their cases of heavy stones or coarse sand (*Fig.* 7.) in contrast to the light vegetable material used by their pond-living relatives, and some, such as *Silo*, even add large anchor stones. The limpet (*Ancylastrum fluviatile*) and *Limnaea pereger*, which is the only snail commonly found in swift-flowing water, have large flat feet, which attach them firmly to the stones (*Fig.* 7). Mayfly nymphs of the family Ecdyonuridae are flattened and apply themselves closely to the stones, onto which the current holds them (*Fig.* 5). In the genus *Rhithrogena* the broad flattened gills are spread out round the body like a flange, and the anterior pair are bent round under the thorax and closely fitted to the substratum. This device enables the nymphs to cling very tightly to smooth surfaces, and they are very difficult to dislodge even from a sheet of glass. Many animals, including some of the caseless caddis-worms, use silken threads for anchoring themselves, and the larva of the blackfly *Simulium* (*Fig.* 8) has a most remarkable method of attachment. It may be noted in passing that the Americans call this insect the buffalo-gnat, a name which will be adopted here as it is more descriptive than the English one. The adult fly does resemble a minute buffalo with huge hunched shoulders and it has nothing to do with the 'blackfly' of the kitchen garden. The larva of *Simulium* spins a small mat of silken threads on some solid object, and then moves its broad backside on to the mat and engages a complex circlet of little hooks in the mesh. It also provides itself with a safety-line, rather in the manner of a spider, so that if it does break free it can clamber back again to its original position. Other insects are provided with stout claws, with which they can cling to small irregularities on stones; these are found on caddis-worms of both cased and caseless types, on the stoneflies, on the little sluggish beetles of the family *Helmidae* (*Fig.* 6), on midge larvae (e.g. the chironomid *Diamesa*) and on mayflies such as *Baetis* (*Fig.* 5). The last is also an active swimmer and is clearly streamlined and able, like the similarly streamlined trout, to swim against quite fast currents. Like the trout also, it generally does so only in furious bursts. The maggot of *Limnophora* has long fleshy hooks in its tail (*Fig.* 8) which presumably help it to maintain its position.

The fishes have already been mentioned in Chapter II. They are either active swimmers, who can hold their places in turbulent water and shelter behind large stones or under the banks, or they have flattened bellies and live on the bottom down amongst the stones, where they are out of the main force of the current. The first group includes the trout and salmon, and in quieter water the grayling and minnow, and the second the bullhead and the stoneloach. Both the last two fishes are confined to places where large stones provide them with shelter, and are absent from areas of clean-swept gravel, where there are no crevices in which to hide

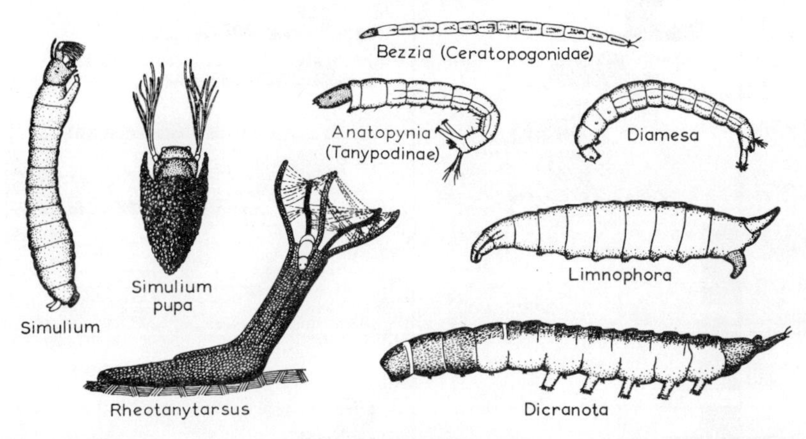

FIG. 8. Dipteran larvae, and the pupa of *Simulium*, of eroding substrata in running water. Magnified x 5. *Simulium* species occur also in weed beds in all types of running water, and larvae closely resembling those of *Bezzia*, *Anatypnia* and *Diamesa* occur in many types of habitat.

from the current. All these species of fish spawn in gravel or amongst stones.

Most of the inhabitants of eroding substrata are cold-loving animals, and cannot tolerate very warm water. Some of them, such as the flatworms *Crenobia alpina* and *Polycelis felina* (*Fig.* 5), can withstand very little rise in temperature, and so are confined to high or cool streams; and others, such as many of the stoneflies, mayflies and caddisflies, get over this problem by doing all their growth in the winter. They then emerge as adults in the spring and their offspring avoid high summer temperatures by remaining as unhatched eggs. The life histories of a typical stonefly and a typical mayfly are illustrated in *Fig.* 9, which is based on an actual investigation.

Four main types of feeding habit are found. Some creatures, as in all animal communities, are carnivores; such are the fishes, the large stoneflies, caddis-worms such as *Plectrocnemia* and *Rhyacophila*, many

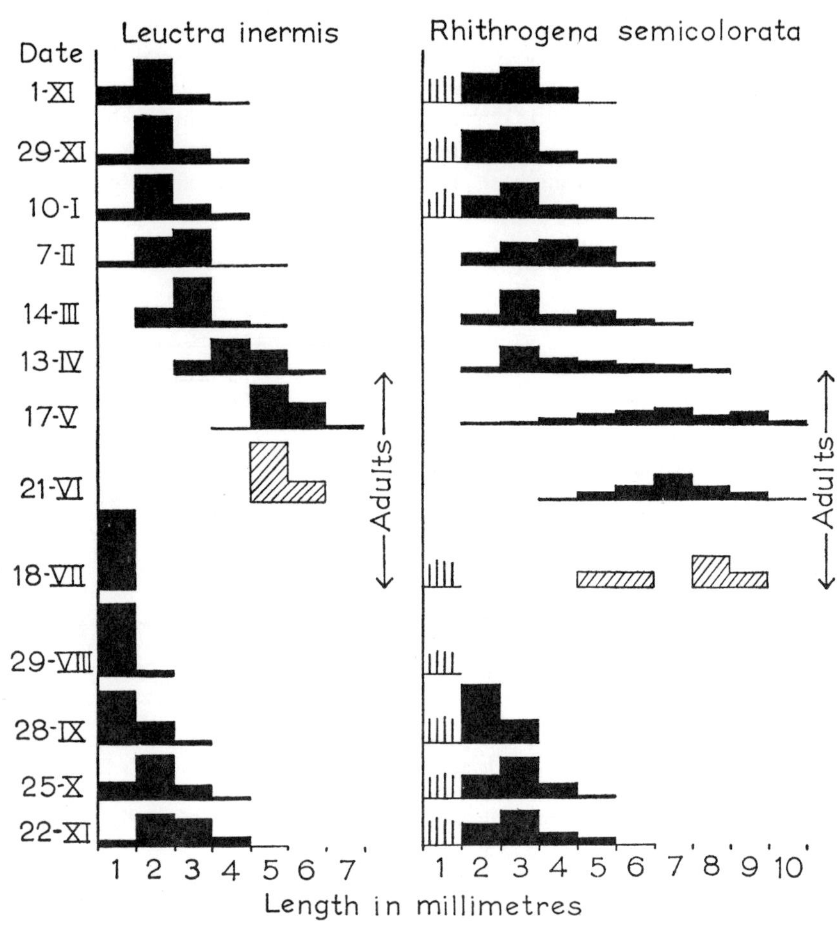

FIG. 9. Diagram illustrating the life cycles of a typical stonefly and a typical mayfly. Based on collections from the Afon Hirnant 1955–6. Each Histogram shows the percentage distribution in each length group of each sample of nymphs. Histograms based on five or fewer specimens are hatched. These small samples occurred during the flight-period of the adults, which is shown by a vertical line between arrowheads, and which is the period during which the eggs were laid. The histograms for *R. semicolorata* are based on samples taken with a coarse net, which did not retain the smallest specimens, but the presence of these is marked by vertical hatching.

beetle larvae and the leeches. These make up a small proportion of the fauna; Percival and Whitehead (1929) found this to be between 1 and 6 per cent., but their figures are undoubtedly too high, for reasons which will be apparent later. The molluscs, most of the mayfly nymphs and many caddis-worms scrape algae off the stones, and Moon (1939) suggests that algae are the main foodstuff of the animals of eroding substrata. Detritus of terrestrial vegetable origin is, however, of considerable importance, and many of the small stonefly nymphs (Hynes, 1941) and some caddis-worms (Percival and Whitehead, 1929) always feed on dead leaves and detritus, as do the shrimps which occur in most of the harder waters where the current is not too swift. Indeed Jones (1950) has suggested that detritus and leaves may well be more important as the basic food of the animal community than algae, whose numbers are liable to violent fluctuations, particularly as the result of floods. The fourth type of feeding habit depends on the current, which always carries with it small particles of detritus, detached algae and animals which have lost their grip or have been swept in from ponds. This source of food is exploited in various ways. The larva of *Simulium*, which hangs out into the water from its silken mat, has a pair of large collapsible comb-like structures on its head, with which it sifts the water by making continuous grasping movements. It thus catches small particles carried by the current, very much as a barnacle on the sea-shore sifts small creatures out of the waves. The small midge larva of the genus *Rheotanytarsus* builds itself a tube on the surface of a stone, from which radiate a number of oblique struts, making the tube resemble a small *Hydra*. The larva then spins a string of sticky saliva between the struts, turning the structure into a little net supported by poles (*Fig.* 8), and this filters the food from the water (Walshe, 1950). Perhaps even more striking are the nets made by several of the caseless caddis-worms, of which the genera *Hydropsyche*, *Philopotamus*, *Plectrocnemia* and *Polycentropus* are examples (*Fig.* 7). These insects spin conical silken nets which are directed upstream and attached along one side to a stone, usually on the underside at a point where it is raised off the bottom. The force of the current keeps the net open, in the same way as the wind keeps open the wind-sock on an airfield, and the larva, which lives in the tail-end of the net, merely cleans its surface at intervals (Wesenberg-Lund, 1943). These current-feeding organisms are common in all types of stony streams, and also in weed beds; wherever, in fact, they can attach themselves and operate their fishing tackle. They become particularly abundant where the water is rich in suspended foodstuffs, and are often found in enormous numbers in the outflows of lakes, where the river water contains masses of plankton derived from the lake (Illies, 1956).

The inhabitants of 'depositing' substrata are less varied and interesting, and, apart from a few carnivores, such as the fishes and the alder-fly *Sialis*, they are scavengers (*Fig.* 10). Silt and mud are unsuitable substrata for algae, and the main foodstuff available is the silt itself, dead leaves and other decaying organic matter. In hot weather, because of the sluggish current, the percentage saturation with oxygen may fall at times to quite low levels, particularly at night, and the habitat is available only to those species which are able to tolerate such conditions. The normal inhabitants are, in fact, nearly all creatures which also live in stagnant water, such as lakes or ponds. On the surface of the mud one finds the water-slater *Asellus*, which feeds on dead leaves and bacterial growths; and in the mud, burrowing creatures and sometimes large numbers of little bean-shrimps (Ostracoda, e.g. *Cypris*) occur. The burrowers include segmented worms, particularly those of the family Tubificidae, which feed like earthworms on the mud itself. Their blood, like that of the earthworms, contains the red pigment haemoglobin, which is an efficient collector of oxygen. The Tubificidae have the curious habit of sticking their tails up into the water above the mud and waving them rhythmically back and forth. When the oxygen content of the water is low there is none at all in the mud, so the only oxygen available to the worms is collected by the blood in their tails and pumped down to their front ends. The lower the oxygen content of the water the further they stick out their tails (Wesenberg-Lund, 1939). Also to be found in such places are the large mussels *Anodonta* and *Unio*, although the latter also occurs in sand and fine gravel, and the pea-mussel, *Pisidium*. These live just below the mud surface, with part of their shell exposed; from this part the two siphons project, and a continuous current of water is passed in at one siphon and out at the other. The siphons of *Pisidium* are relatively very long, and the animal itself is often deeply buried so that only the tips of the siphons project above the mud. Other molluscs which occur commonly in river muds are the operculate snails *Bithynia* and *Valvata*, which plough along on the surface and feed on detritus. Another important group of burrowing animals are the larvae of several species of the midge *Chironomus*, which like the Tubificidae have red blood; hence their popular name 'blood-worms'. They make tubes in the mud and feed on particles of silt, bacteria and detritus on the surface. Most species drag these into their burrows with strings of saliva which they spread out on the surface, but one common species, *C. plumosus*, spins a conical net of saliva in the mouth of its burrow and draws in a current of water by undulating its body in the tube. This sweeps in small particles which are caught by the net, and the larva eats and replaces the net at intervals (Walshe, 1951). It therefore behaves rather

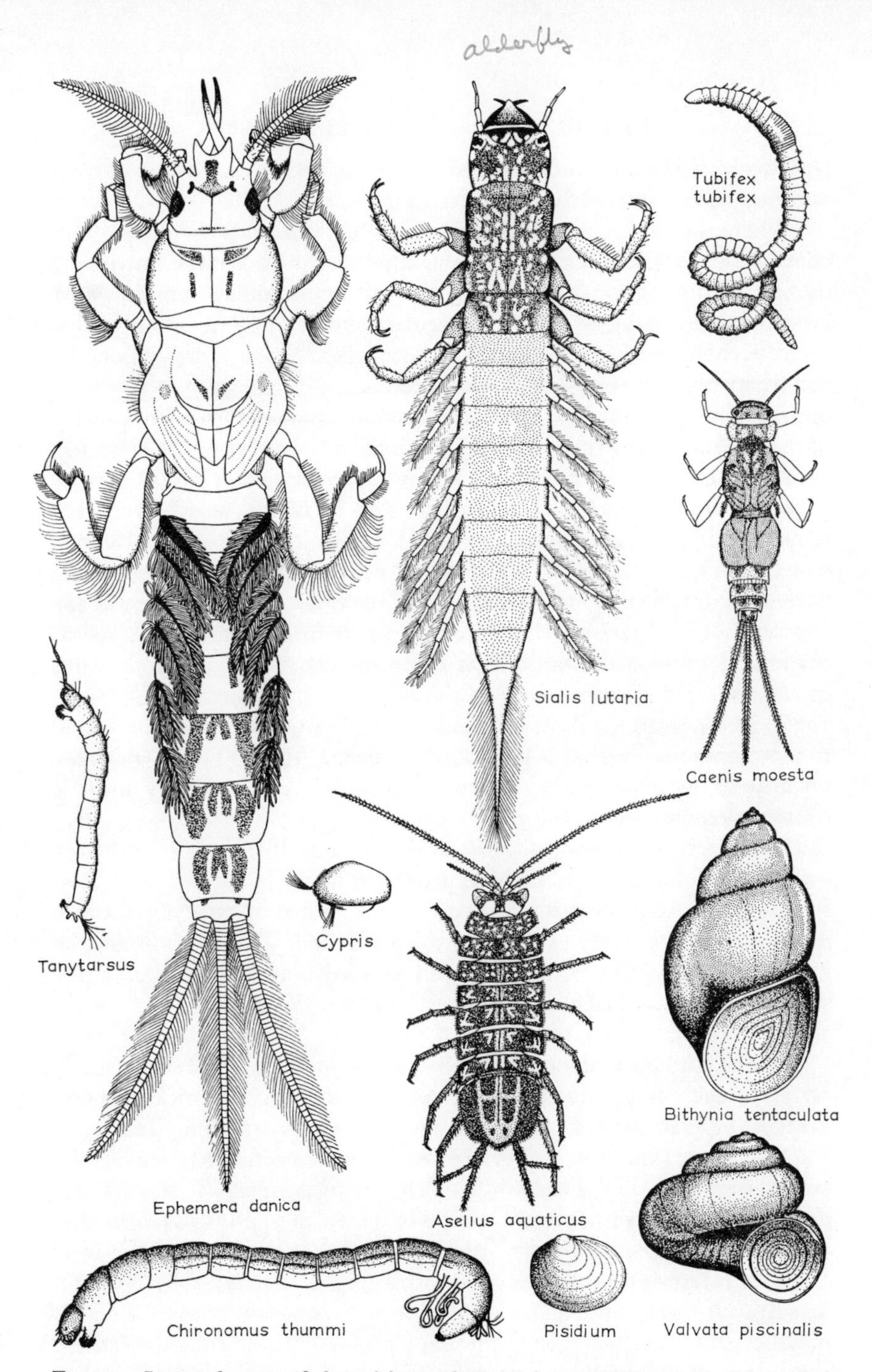

FIG. 10. Invertebrates of depositing substrata in running water. Magnified x 5. All these animals may also occur on or in similar substrata in lakes and ponds.

like the net-spinning caddis-worms except that it makes its own current and eats its net instead of merely cleaning it.

Where the silt is coarse rather than muddy, one finds also a few other characteristic creatures. Most notable amongst these are the burrowing nymphs of the fisherman's mayfly *Ephemera*, and the small tube-building larvae of *Tanytarsus*, which often occur in such enormous numbers that the river bed seems to be made of their tubes. Great numbers of tiny mayfly nymphs of the genus *Caenis* are also often present wherever silt occurs. The two mayflies which are characteristic of this habitat are of particular interest as they show adaptations to their way of life (*Fig.* 10). Most mayfly nymphs create a current of water over their bodies by flapping their gills. An exception to this is *Baetis*, which does not move its gills; probably because, being an inhabitant of turbulent water, it needs to make no respiratory current of its own. In most genera the current is drawn in from the sides and then flows down the back of the nymph, but in *Ephemera* the large fluffy gills work together as a membrane and cause a current which flows directly along the body, thus causing no disturbance to the walls of the narrow burrow (Eastham, 1939). In *Caenis* the gills of each side of the body beat out of phase, and thus a transverse current is produced (Eastham, 1938). This enables the creature to live and breathe in silty places, as it has to keep only one side of its abdomen clear of the silt.

Most of the burrowing animals of depositing substrata live in the top inch or two, but the Tubificidae penetrate down to depths of at least four inches, and it has been shown recently that they raise a great deal of mud to the surface, as do earthworms in soil. This was discovered by use of radioactive material spread in a layer two inches below the surface of mud to which Tubificids were added (Westlake and Edwards, 1957).

All the animals of depositing substrata mentioned above live actually on or in the silt or mud; but on solid objects such as sticks, stones, bridges, or even the old prams and bicycles which are such a feature of English rivers, one finds other creatures which need shelter or a solid substratum (*Fig.* 11). These include the sponges *Spongilla* and *Euphydatia*, the polyp *Hydra*, several species of flatworm, such as *Dendrocoelum* and *Polycelis nigra*, the moss animalcule *Plumatella*, leeches, the snail *Limnaea pereger* and some caddis-worms (e.g. *Anabolia*). One does not usually, however, find many of the normal stone-dwelling species of mayflies and stoneflies, nor the limpet: probably these animals find the conditions of temperature, oxygen régime and general siltiness intolerable.

Where macrophytes are present the amount of shelter is greatly increased and the fauna is correspondingly enriched. This applies not only to the invertebrates but also to the fishes, because almost all the fishes of the slower reaches of rivers, which include the pike, perch, roach, rudd, chub, bream and dace, require weeds on which to spawn. The

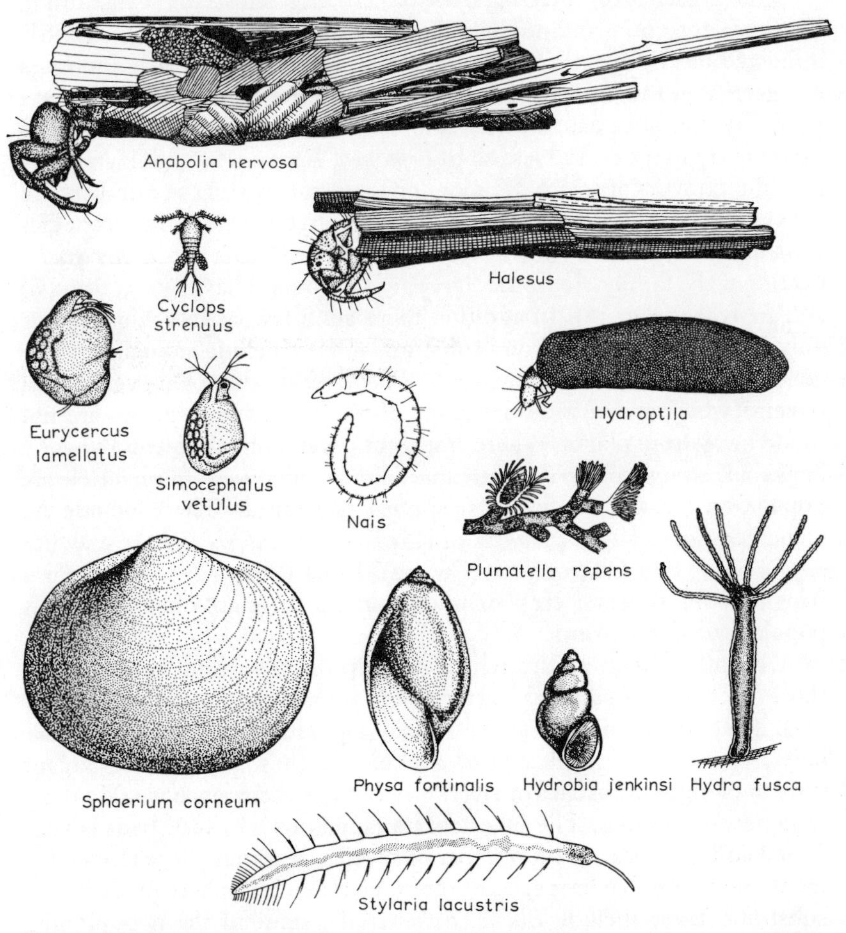

FIG. 11. Polyps, moss-animalcules, segmented worms, crustaceans, caddis-worms and molluscs of solid objects and weed beds of depositing subtrata in running water. Magnified x 5. All these animals may also occur in lakes and ponds, and some may also occur on eroding substrata where the current is slack.

three-spined stickleback is the only British freshwater fish that breeds on a weedless muddy bottom, and other fishes which are found in weedless reaches are only temporary inhabitants which have moved in to feed. Fishes are, in fact, very mobile and often travel long distances, and this applies to all species, not only to the salmon, sea-trout and eel, which are well known to migrate. For this reason they are, despite their popular reputation in the matter, very poor indicators of river conditions: in winter particularly they often move into badly polluted water.

The presence of macrophytes on eroding substrata, constituting Butcher's torrential and non-silted communities (Chapter III), has little influence on the *composition* of the fauna, although it greatly increases the *density* of the animals, because many species of worms and insects profit by the abundant shelter and food in the plants. This applies particularly to moss, and less so to *Lemanea*, *Ranunculus* and *Myriophyllum*, the growths of which are more open. Creatures whose numbers are always increased by the presence of macrophytes include the stoneflies *Amphinemura*, *Leuctra* and *Isoperla*, the mayflies *Baetis* and *Ephemerella*, beetles of the family Helmidae, larvae of Chironomidae (*Figs.* 5, 6 and 8) and *Ephydra* (*Fig.* 12). In addition there are a few species which occur only in macrophytes on stony substrata—these include stoneflies of the genus *Protonemura* and the mayfly *Baetis niger*; except in very small streams which are outside the scope of this book these species are not found away from plants. Where emergent vegetation is present along the banks of stony rivers and streams similar increases in invertebrate population occur, and a few extra species may appear. These include the stoneflies *Taeniopteryx nebulosa* and *Nemoura avicularis* and the mayflies *Leptophlebia* and *Centroptilum* (*Fig.* 12). In hard water where there is a fringing bank of water cress or water parsnip there are often enormous populations of *Gammarus*.

The plant communities which develop on depositing substrata are those we have classified as partly silted, silted, and emergent (Chapter III). These profoundly alter the animal population, as they provide not only a greatly increased area of shelter but they allow the abundant growth of algae in reaches of rivers where their occurrence would otherwise be very limited. The invertebrate fauna of such weed beds is very varied and includes a great number of species, of which only a few can be mentioned here. Some species of the stone fauna are able to find suitable substrata: these include *Baetis*, *Ephemerella*, some of the net-spinning caddis-worms and *Simulium*. Weeds also provide habitats for many species which can tolerate low oxygen contents at times, but which are unable to live in bare mud or silt. Among these are the mayflies *Leptophlebia* and *Centroptilum*, the dragonflies *Agrion* and *Ishnura*, the snail *Physa fontinalis* and the pea-mussel *Sphaerium corneum*, which clambers about on the plants with its long tongue-like foot. One finds also those animals, already mentioned, which occur on sticks and stones in silted reaches, many caddis-worms which make their cases out of vegetable

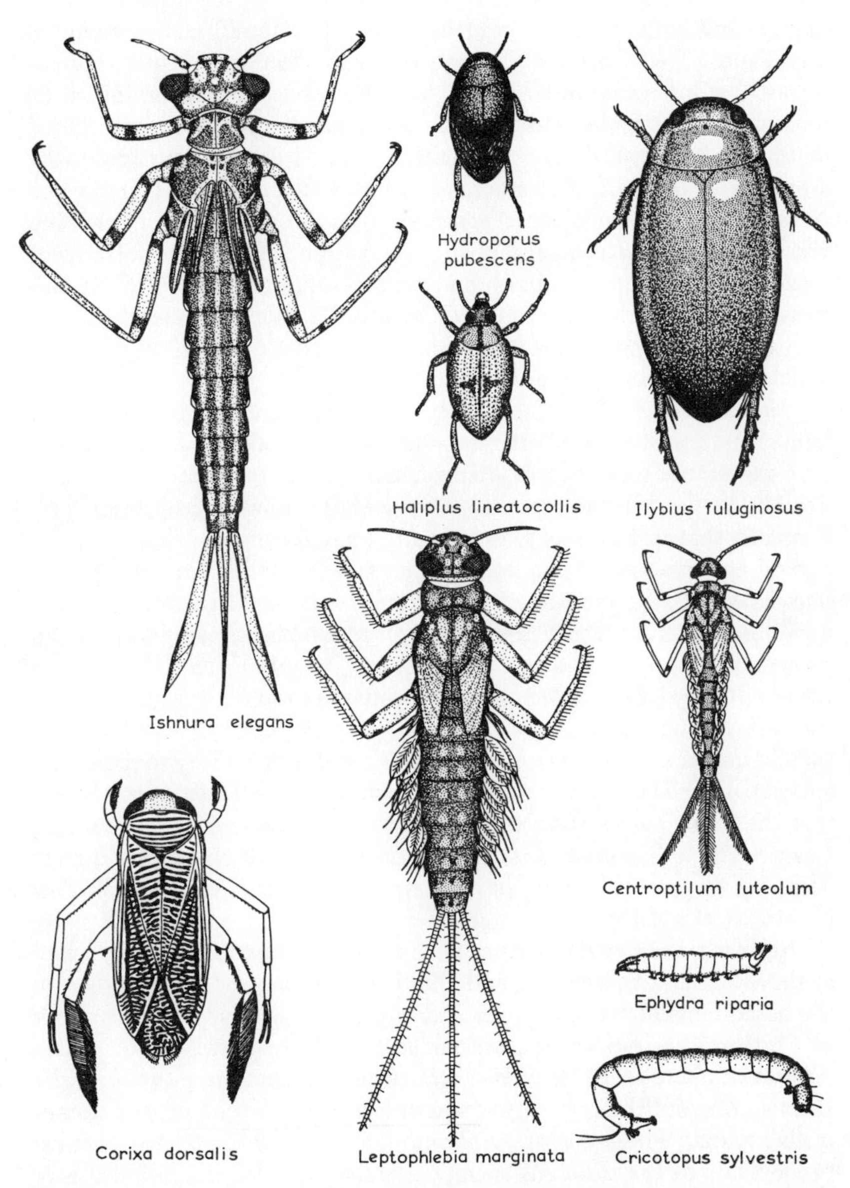

FIG. 12. Mayfly and dragonfly nymphs, water-boatmen, beetles and fly larvae of weed beds on depositing substrata in rivers, lakes and ponds. Magnified x 5.

matter, and large numbers of small green chironomid larvae. Some of these, and a few of the caddis-worms, actually feed on the weeds themselves, but others merely burrow into the leaves and there spin little filter-nets to catch the flotsam as does *Chironomus plumosus* in the mud below. In the dense shelter of the littoral vegetation one may find many creatures of the truly pond fauna such as water boatmen (*Corixa* and *Notonecta*), water scorpions (*Nepa cinerea*) and many species of mites and water beetles (e.g. *Ilybius* and *Hydroporus*). Some of these may also occur in the weed beds, particularly the beetle *Haliplus*. Several of these animals are illustrated in *Figs.* 11 and 12, and the above list could be very considerably lengthened without including all the kinds of creatures which are commonly encountered.

So far we have been dealing only with the general composition of the fauna. Its density (numbers per unit area) and detailed composition vary under the influence of many factors, important among which is liability to flooding and periods of torrential flow. It has often been observed that floods sweep away many animals and sessile algae, and several workers have demonstrated that severe floods have reduced the population of invertebrates. Jones (1951b) reports that severe summer floods in the river Towy reduced the invertebrate population from 300–1,000 per quarter square metre to 40–48, and he (1958) observed a similar but less severe effect of a spring flood in the river Ystwyth. This sort of disaster also, of course, influences the normal density of the population, which has been shown to vary with the liability to damaging spates. In the Tees, which suffers from sudden high floods, it was found that the fauna varies from just over 100 to over 800/sq. metre at different stations (Butcher *et al.*, 1937), whereas in the more placid river Avon up to 3,000 or more animals were found in a comparable area (Pentelow *et al.*, 1938).

Similarly the detailed nature of the substratum controls the density of the invertebrates. Percival and Whitehead in their classical studies on the inhabitants of streams (1929 and 1930) showed that this is a matter of considerable importance, as can be seen from some of their data reproduced here as Table 2. It is clear from this that the more fixed the substratum, and the greater the amount of shelter available, the denser is the fauna. The type of substratum also influences the numerical proportions of the different species. Percival and Whitehead also give a table, reproduced here in a simplified form (Table 3), showing the percentage composition of the fauna found on various types of stream bed. From this it can be seen that some creatures, such as *Baetis*, *Rhithrogena*, *Glossosoma* and *Agapetus* are much commoner on stones than they are in vegetation, and that others, such as Chironomidae,

TABLE 2

Number of invertebrates per square metre found on different types of stream bed in the rivers Aire, Nidd and Wharfe, Yorkshire. Data from Percival and Whitehead, 1929.

Loose stones	3,316
Stones embedded in the bottom	4,600
Small stones with fine gravel	3,375
Blanket-weed on stones	44,383
Loose moss	79,782
Thick moss	441,941
Pond weed (*Potamogeton*) on stones	243,979

Limnaea and *Gammarus*, are particularly associated with plants. The table also shows that some animals are particularly prevalent on one type of substratum, e.g. *Rhithrogena* on loose stones and gravel, *Leuctra* on larger stones and *Hydropsyche* in thick moss. Here we are meeting the importance of the microhabitat; all the habitats shown in Table 3 belong to the eroding zones of streams, and yet within this one environment small differences in substratum result in large differences in the proportions and to some extent in the kinds of the different species present. Another example of this effect is given by Jones (1949a), who found that the fauna of the Welsh river Sawdde is more limited in number of species than is that of its tributary the Clydach merely because the slope of the Sawdde is more even and the bottom more uniform. It therefore has a greater uniformity of microhabitat and so offers suitable conditions for fewer species; in other respects the two rivers resemble one another closely.

Whitehead (1935) also made a study of a stream at Driffield, Yorkshire, and his findings are shown in simplified form in Table 4. This illustrates the same differences between the stony bed and weeds, but it also shows very clearly the differences between the relatively soft and rapidly flowing rivers he had investigated earlier (Table 3) and the calcareous less rapid Driffield stream. In the latter, molluscs other than *Ancylastrum*, and *Gammarus* were of far greater importance. These are, as we have seen, all species favoured by hard water, and *Gammarus* is always particularly abundant where it is not likely to be swept away by swift currents or sudden floods.

The differences between the faunas of the eroding reaches of several rivers are shown in Table 5, where the rivers are arranged in order of

TABLE 3

The average percentage composition, to the nearest whole number, of the chief types of fauna from different types of stream bed. Data taken from Percival and Whitehead (1929) *and somewhat simplified.* * = *less than* 0·5 *per cent.*

Type of stream bed:	loose stones	stones embedded in gravel	small stones & small gravel	*Cladophora* on stones	loose moss on stones	thick moss	Weed on stones
Worms							
Naididae	1	1	8	12	4	33	—
Crustacea							
Gammarus	1	—	1	*	1	1	4
Stoneflies							
Amphinemura	1	*	—	*	*	*	—
Leuctra	2	2	—	*	*	*	—
Isoperla	1	*	1	*	*	*	*
Perla	*	—	—	—	*	—	—
Chloroperla	*	—	—	—	—	—	—
Mayflies							
Baetis	15	5	2	2	5	2	2
Caenis	*	—	2	1	2	*	—
Ephemerella	1	9	—	9	11	5	5
Ecdyonurus	7	4	1	*	*	*	—
Rhithrogena	12	—	8	*	*	*	—
Caddis-worms							
Rhyacophila	1	*	—	1	*	*	*
Polycentropus	1	3	—	1	*	*	1
Hydropsyche	1	—	—	1	*	4	—
Glossosoma	3	4	3	*	—	—	—
Agapetus	20	21	18	2	*	—	—
Psychomyia	*	3	1	3	—	*	—
Hydroptila	—	*	—	5	2	*	—
Ithytrichia	—	*	—	*	—	*	—
Crunoecia	1	1	1	*	2	*	—
Beetles							
Helmidae	3	8	34	8	6	4	—
Midges							
Chironimidae	17	11	5	40	54	41	42
Buffalo-gnats							
Simulium	3	—	—	*	1	2	1
Mites							
Hydracarina	2	2	1	4	3	3	11
Snails							
Limnaea	—	4	1	2	*	3	17
Limpets							
Ancylastrum	—	11	3	3	1	—	*

decreasing water hardness. It can be seen that as the hardness of the water declines a number of changes occur. Worms, shrimps, molluscs and finally chironomids tend to decline in importance, to be replaced by various insects, particularly mayflies; and stoneflies increase steadily in

TABLE 4

*The percentage composition, to the nearest whole number, of the fauna of three types of habitat in Great Driffield stream, Yorks. Data taken from Whitehead (1935) and considerably simplified. Only those animals are shown which made up at least 0·5 per cent. of the fauna in one habitat. * = less than 0·5 per cent.*

Type of habitat:	stony bed	water-parsnip *Sium erectum*	Mares-tail *Hippuris vulgaris*
Flatworms			
Tricladida	1	2	2
Worms			
Oligochaeta	27	12	7
Leeches			
Hirudinea	1	1	2
Crustaceans			
Gammarus	31	28	14
Asellus	*	1	1
Mayflies			
Baetis	3	9	18
Ephemerella	*	1	7
Caddis-worms			
Agapetus	26	1	1
Silo	1	—	—
Limnephilidae	3	2	1
Hydroptila	—	—	1
Beetles			
Helmidae	1	1	*
Others	1	*	*
Craneflies			
Tipulidae	1	*	—
Midges			
Chironomidae	2	9	16
Ceratopogonidae	*	*	1
Snails			
Gastropoda: part	*	1	24
Limpets			
Ancylastrum	*	1	*
No. of animals in samples	3,394	3,164	3,876

TABLE 5

*The percentage composition of the faunas of eroding substrata in various British rivers, arranged, from left to right, in order of decreasing water hardness. Some of the data are taken from published work as indicated. Figures to nearest whole number. * = less than 0·5 per cent.*

	Great Driffield, Yorks., gravel (Whitehead, 1935)	R. Avon, Glouc. (Pentelow *et. al.*, 1938)	R. Dee, Denbighshire	R. Ceiriog, Denbighshire	Trout stream, Northumberland	R. Towy, Cardiganshire (Jones, 1951b)	Afon Hirnant, Merioneth
Season of investigation:	all	all	winter and spring	winter and spring	spring and summer	summer	all
Flatworms							
Tricladida	1	—	1	*	2	—	*
Worms							
Oligochaeta	27	47	8	4	1	1	*
Leeches							
Hirudinea	1	*	1	*	—	—	—
Crustacea							
Shrimps (Amphipoda)	32	3	1	1	—	—	—
Asellus	*	*	—	—	—	—	—
Mayflies							
Baetis	3	1	24	15	22	26	5
Caenis	—	4	1	—	*	—	—
Ephemerella	*	—	22	6	2	19	4
Ephemera	—	*	—	—	—	—	—
Ecdyonuridae	—	—	6	17	15	1	19
Stoneflies							
Leuctridae and Nemouridae	—	—	4	14	23	9	47
Perlidae and Perlodidae	—	—	1	2	4	*	2
Chloroperla	—	—	*	2	1	—	2
Caddis-worms							
Caseless	*	1	3	6	3	*	9
with cases	30	2	*	*	*	5	*
Bugs							
Hemiptera	—	1	—	—	—	*	—
Beetles							
Helmidae	1	2	3	1	6	8	1
Other beetles	1	—	*	—	*	2	1
Midges							
Chironomidae	2	22	7	33	16	5	4
Ceratopogonidae	*	—	1	—	1	*	—
Buffalo-gnats							
Simulium	*	—	—	—	1		6
Other flies							
Diptera	1	—	—	—	—	1	*
Water-mites							
Hydracarina	*	—	1	—	*	23	—
Snails							
Gastropoda	*	16	1	—	1	—	—
Limpets							
Ancylastrum	*	1	14	*	2	*	*
Other animals							
(mostly insects)	1	—	—	—	—	—	*
Total no. in collections	3,372	3,258	4,444	1,620	13,745	2,657	7,788

importance. These are of course only general trends, which are to some extent obscured by individual differences between the rivers from which the data were obtained. It is, however, a general truth that in very soft waters stoneflies and mayflies are the dominant creatures; in slightly harder waters mayflies and chironomids become more dominant; and in very hard water the dominants are usually shrimps, snails, worms and caddis-worms with cases.

Even within one water course, whose general character does not alter as one proceeds downstream, trends in the abundance of certain species are often found. Table 6 shows some figures obtained from the Afon

TABLE 6

The change in abundance of certain species in the Afon Hirnant with altitude. The numbers are percentage composition of the total fauna at each station

Altitude of station in feet:	700	800	950	1,100
Baetis spp.	4·6	9·2	10·1	11·7
Ephemerella ignita	3·8	4·1	4·8	5·4
Rhithrogena semicolorata	17·6	11·7	6·9	2·7
Heptagenia lateralis	0·1	0·1	1·0	1·1
Amphinemura sulcicollis	24·1	23·1	25·8	7·6
Isoperla grammatica	1·0	3·3	5·3	11·6
Hydropsyche instabilis	7·3	1·8	0·2	<0·1
Stenophylax spp.	0·1	0·2	0·2	0·5
Helmis maugei	0·3	1·3	2·8	12·7
Orthocladiinae	2·5	3·5	3·6	7·4

Hirnant, a typical mountain trout-stream in North Wales. Apart from the fact that it increases in size in the 3½ miles of its course from 1,100 feet to 700 feet above sea level its general character remains unaltered, and the only important change seems to be that there are no trees above 1,000 feet. Nevertheless there are marked changes in the abundance of several important constituents of the fauna. Some of these trends can be explained on the basis of food supply. For example *Amphinemura* feeds very largely on dead leaves, and these are more abundant where there are trees. Similarly the net-spinning caddis *Hydropschye* presumably finds more to eat in water that has travelled further and had more time to pick up algae and small animals. Liability to flooding, which increases as one proceeds downstream, probably accounts for the reduction in *Baetis*, which relies on its claws and swimming powers to resist the current, and its replacement by *Rhithrogena*, which has an efficient hold-fast mechanism. These two genera eat similar food, and so may compete directly with one another. It may be that the trends in *Helmis, Isoperla, Stenophylax* and the Orthocladiine chironomids can also be explained in terms of flooding. The main point about such trends, for our purpose here, is to note that most of them are gradual, some are upwards and some downwards, and some creatures do not show them. They are therefore different from the sudden changes affecting the entire fauna, which are, as we shall see, often caused by pollution.

All the examples of river faunas given above have been based on surveys which have covered all or most of the year, and so have included all the animals. It has, however, already been pointed out that many insects avoid the high summer temperatures by emerging as adults in the spring or early summer, their offspring passing the warm season as eggs. Others do just the reverse, so the composition of the fauna undergoes marked seasonal changes. An example of this is shown in Table 7

TABLE 7

*The percentage composition of the fauna of the Afon Hirnant at four stations each month throughout a year. Only those species are shown which made up at least 2 per cent. of the fauna in one month. Figures are shown to the nearest whole number. * = less than 0·5 per cent.*

Date of sampling:	1 XI	29 XI	10 I	7 II	14 III	13 IV	17 V	21 VI	18 VII	29 VIII	28 IX	25 X	22 XI
Mayflies													
Baetis spp.	7	12	10	11	4	4	5	12	9	24	9	7	10
Ephemerella ignita	—	—	—	—	—	—	*	29	36	9	*	*	*
Ecdyonurus venosus	6	3	1	1	*	*	1	1	1	1	5	2	1
Rhithrogena semicolorata	16	14	16	12	12	10	13	2	*	—	4	10	8
Heptagenia lateralis	*	—	*	*	1	1	3	2	1	*	—	—	*
Stoneflies													
Brachyptera risi	—	*	4	7	3	2	—	—	—	—	—	—	*
Protonemura spp.	7	6	5	3	*	*	—	*	1	5	9	5	4
Amphinemura sulcicollis	24	27	27	26	22	11	—	1	*	—	20	39	32
Leuctra hippopus	2	1	*	*	*	*	—	—	—	*	2	2	1
Leuctra inermis	7	15	18	18	33	30	25	3	—	—	3	15	23
Leuctra fusca	—	—	—	—	—	—	—	3	13	11	1	*	—
Isoperla grammatica	3	8	7	8	7	8	8	3	1	*	1	1	3
Chloroperla spp.	1	1	1	1	2	3	5	3	*	1	2	2	3
Caddis-worms													
Rhyacophila spp.	1	1	*	1	1	1	1	2	2	5	3	1	1
Plectrocnemia conspersa	2	1	*	*	*	1	2	2	1	2	2	*	*
Hydropsyche instabilis	5	2	2	2	5	3	4	4	1	1	1	3	1
Beetles													
Helmis maugei	7	3	1	2	5	10	4	3	8	7	14	1	1
Midges													
Orthocladiinae	5	*	*	*	1	11	9	12	4	4	11	3	2
Tanypodinae	*	*	*	—	—	1	*	2	5	*	*	*	*
Buffalo-gnats													
Simulium spp.	*	2	5	5	1	1	5	15	13	23	5	2	8
Total no. in sample	2,380	3,543	3,846	2,885	1,371	3,438	1,743	2,218	1,811	580	575	1,671	4,367

and *Fig.* 13 where data from the Afon Hirnant are set out month by month as they were collected. The table shows that some species such as

Amphinemura sulcicollis, *Leuctra inermis* and *Rhithrogena semicolorata*, which are very important members of the fauna in the winter and spring, were almost or quite absent for several summer months, when their place was taken by *Leuctra fusca* and *Ephemerella ignita*. The table is

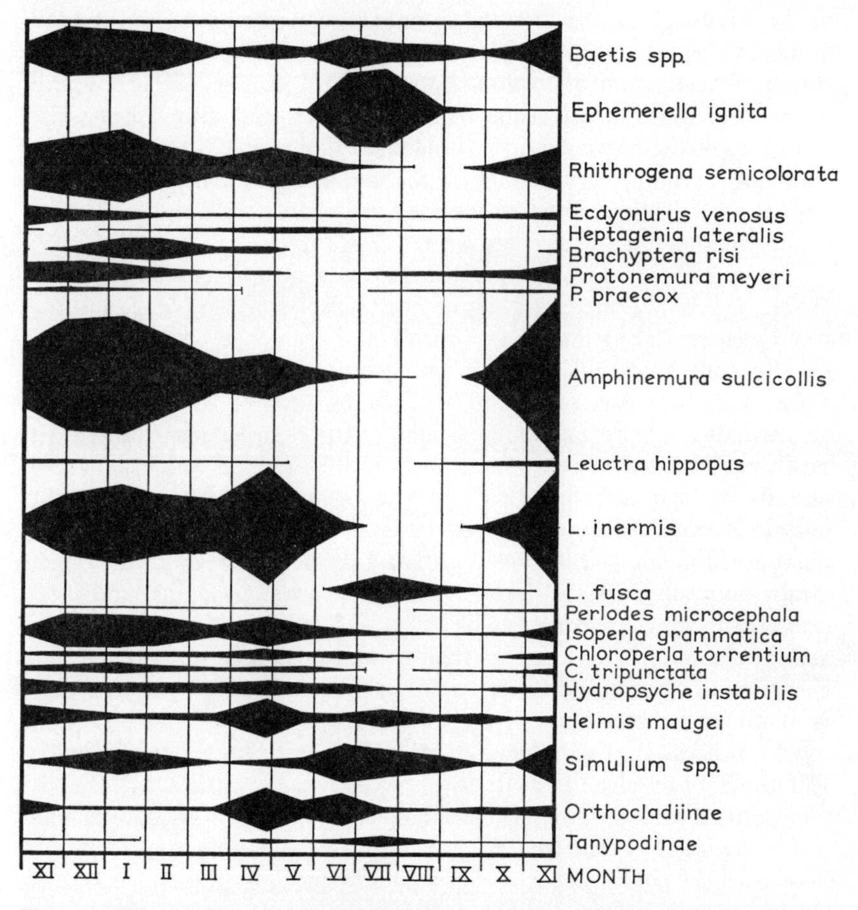

FIG. 13. The numbers of animals caught month by month during 1955–6 in the Afon Hirnant by a standardised netting technique. The width of each Kite-diagram at each sampling period is proportional to the number of specimens of that species caught. Compare with TABLE 7.

based on percentage composition, which brings out the changes in proportions, but the figure is based on the actual numbers collected and this shows other things. Firstly it emphasises the irregularity in total numbers which were collected by the sampling technique: for instance the

March sample was a poor one, because of flood conditions, and the April sample was particularly large because of very low water. This sort of thing is unfortunately unavoidable, but, as can be seen from the table, it has little effect on the determination of the composition of the fauna. Secondly, the figure shows that although some species do fill in the gap in the summer months they do so only to a limited extent: the total numbers present in the summer are much lower than in the winter or spring. Investigation of trout-streams, and to some extent of lowland rivers also, must therefore be made during at least two seasons if a good idea of their fauna is to be obtained. The least satisfactory season is the late autumn, when almost all the insects are small and difficult to find. If investigations must be confined to a short period the most satisfactory seasons for this are late winter or spring.

There remains the all-important question of the method of investigation. Inevitably one can examine only samples of the fauna, and the way these are taken may have considerable influence on the findings. Clearly collections must contain enough creatures; just turning over a few stones or sieving a pint or two of mud is insufficient. It is also an advantage to obtain some idea of the number of creatures per unit area. This is fairly easy on depositing substrata on which some sort of grab can be used to cut out a definite area, and where in any event the bottom is fairly uniform. In weed beds and on eroding substrata it is far more difficult and no really satisfactory solution has been found. Grabs will not work satisfactorily except in rare situations, and most sampling techniques have been based on scoops which take out a definite area of the river bed. This has then to be laboriously sorted over, and the scoops always crush and destroy many of the animals. Another method is to stir up an area of the river bed and catch any creatures dislodged by the process in a net (Davis, 1938). This is in many ways a better technique, but it also has its disadvantages. Many creatures, particularly *Ancylastrum* and *Rhithrogena*, are dislodged only with great difficulty, and some, such as *Rheotanytarsus*, are almost impossible to catch in this way. Also the amount caught is greatly influenced by the rate of flow of the water, which forms eddies in the mouth of the net. Consequently samples taken at one place often yield many more animals at times of low water than they do during floods. After trial of many types of apparatus on many types of rivers I myself have become a devotee of collections made with an ordinary hand-net; not from a definite area, but for a fixed period or in a standardised manner. This suffers from the objections outlined above, and in addition gives only relative, not absolute, differences between collecting stations, but the technique has the merit of great flexibility and adaptability to different situations, and enables one to sample all the types of river bed more or less in proportion to their areas.

Finally there is the question of size of mesh. Whether one sieves or uses a net one employs some sort of a mesh. The size of this can greatly influence results, because it is generally true that any creature which can get through a net will do so. Jónasson (1955) has shown that many conclusions about the life histories of the insect larvae of lake muds have been erroneously arrived at because of the use of too large a mesh size, and the same doubtless applies also to many studies of streams. All

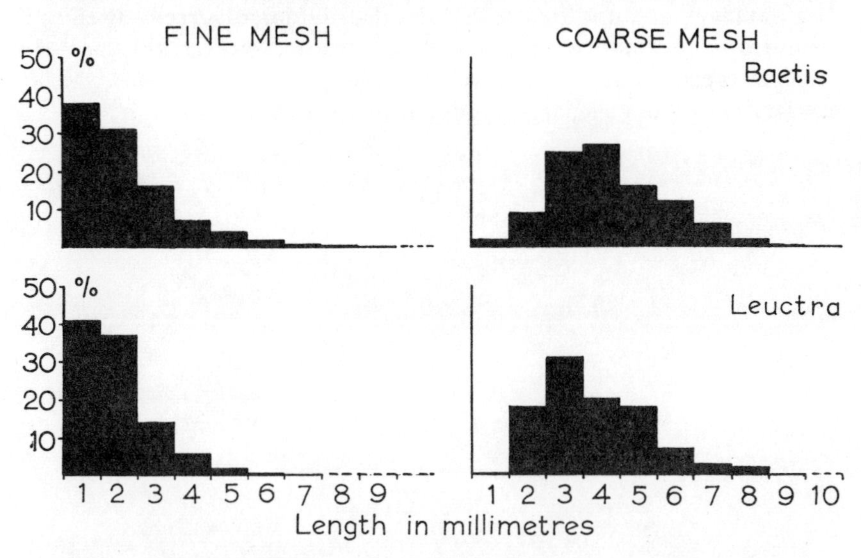

FIG. 14. Histograms demonstrating the difference in percentage size-distribution of samples of *Baetis* and *Leuctra* collected with a net of coarse mesh (30 meshes/in.) and one fine mesh (120 meshes/in.). Figures based on several thousand specimens from the Afon Hirnant 1955–6.

the investigations cited above have been based on collections obtained with fairly coarse nets, which allowed the escape of very young specimens. The use of a very fine mesh raises all sorts of practical problems, such as increase of eddying in the mouth of the net and very tedious and lengthy sorting. However, as will be seen from *Fig.* 14, the usual coarse nets miss all the very tiny specimens, and many of the fairly small ones. That is why it was suggested earlier that Percival and Whitehead's figures of the percentage of carnivores were too high. Carnivores are large and so were caught by their techniques, which, however, lost a large number of the smaller prey organisms. Similarly the figures for the Afon Hirnant, particularly as shown in Table 7 and *Fig.* 13, are to some extent erroneous. I know, on the evidence of fine-net samples, that those species which disappeared for a while reappeared as tiny specimens about a month before they began to appear in the coarse-net samples. For a full discussion of the problems of sampling stony streams the reader is referred to Macan (1958).

Details of technique, however, concern only the investigator, and if he is aware of the difficulties he can to some extent avoid their consequences. As in all ecological work, the biologist has to be content with only approximate numerical results; all sampling techniques have their disadvantages, and it is known that the distribution of stream-dwelling creatures is not uniform (Mottley *et. al.*, 1939). However, this lack of absolute precision in no way invalidates the general conclusions which can be drawn from a suitable study of the inhabitants of a river.

Chapter V

EFFLUENTS AND CHEMISTRY

THE effluents which are discharged into our rivers are as varied as the human activities which produce them. Fortunately, however, for those concerned with the problems of pollution, Southgate (1948) has given a very helpful and informative account of them and of the available methods of treatment. The reader is referred to Dr. Southgate's book or to his more recent summary (1957b) for details, and much useful information is also given by Klein (1957a). Only a very general account will be given here.

Broadly speaking, effluents fall into six groups, but many have to a greater or lesser degree the polluting properties of at least two of these categories.

Inert suspensions of finely divided matter result from many types of mining and quarrying, such as that for china clay, and from washing processes, such as those of coal and root crops.

Poisons in solution occur in the waste waters from many industries. They include acids and alkalis, chromium salts from tanning and electroplating, phenols and cyanides from chemical industries and gas works, tar acids and ammonia, chiefly from gas works; copper, lead and zinc from various industries and mines, insecticides from sheep dips and agriculture, and many others, including the future threat of radioactive isotopes (Wilson, 1957). Some of these substances are fairly rapidly precipitated in the waters into which they flow, as for example are copper and lead when they are run into hard water, but others are very persistent. A few, such as ammonia, are destroyed fairly rapidly by oxidation, while others, such as phenolic and cyanide compounds, are similarly destroyed, but much more slowly.

Inorganic reducing agents, such as sulphides and sulphites, occur as constituents of the effluents of several types of industry. These substances use up the oxygen in river water, and important amongst them are ferrous salts, which are present in many underground waters. These often reach the rivers quite naturally, but large volumes are pumped up by mines and can produce serious pollution. Ferrous waters are usually acid, but as the acid becomes neutralised, usually by the loss of carbon dioxide to the atmosphere, the ferrous salts become oxidised, often by bacterial action, and ferric hydroxide ('ochre') is precipitated. This then takes on the role of an inert suspension. One of the most striking examples of a river affected in this way is the Don, whose bed is rusty red for miles below Sheffield.

Oil is washed into rivers from spills on roads and factory floors, from workshops and garages, and often by the thoughtless motorist who drains the sump of his car and tips the dirty oil into the nearest ditch or stream. This last is a particularly reprehensible habit, as a small quantity of oil can cover an enormous area of water, something of the order of four acres to the gallon.

Organic residues include the effluents from a great variety of activities including dairies, ensilage, manure heaps and cattle yards, slaughter houses, beet-sugar factories, textile manufacture, canning plants, laundries, breweries, tanneries, fish-meal factories, paper mills and, most important of all, domestic sewage. These effluents vary a great deal, but they have much in common. They all contain complex organic compounds in solution and suspension, often together with toxic substances and various salts. Sewage, for instance, unless it be from a very small town, usually contains some gas-works effluent and probably some other industrial wastes; the drainage from silage is usually acid (Rasmussen, 1955), as is that from sugar factories, and paper-mill wastes often contain sulphite. Fundamentally, however, the basic property of these effluents is that they contain unstable compounds which are readily oxidised and so use up the dissolved oxygen in the water. Some of these compounds are more readily decomposed than others: for example dairy wastes oxidise rapidly, but wood pulp is comparatively stable.

Hot water is produced by many industries which use water for cooling purposes, as for instance, steel works and electrical power stations. The water used is usually river water which is pumped through the cooling system, and sometimes raised to very high temperatures during part of its journey. When the original water is polluted, as it often is, trouble is experienced because of growths of bacteria, and it is common practice to chlorinate the water. Usually, unless there is carelessness, the dose is regulated so that none remains as free chlorine when the water is returned to the river. This is possible because chlorine combines readily with organic matter in the water. Unfortunately, however, chlorine combines with the thiocyanates of gas-works effluents to form very toxic compounds (Allen *et al.*, 1946), thus chlorination may produce a poisonous effluent even if no free chlorine remains in it. Moreover, either the heat or the chlorination kills living creatures in the water, so a heated effluent may be not only hot but full of dead organic matter and possibly also poisons.

All these various types of waste water can, of course, be treated in such a way as to reduce their polluting effects, but only in a few instances can treatment be entirely effective. This is because of the law of diminishing returns; it is comparatively easy to remove most of any substance from solution or suspension, but as the concentration falls, removal of more becomes increasingly difficult. Acids and alkalis can be neutralised by the addition of alkalis and acids respectively, but even with such apparently simple treatments difficulties often arise. This is well illustrated by hydrochloric and sulphuric acids. Obviously the cheapest and most convenient alkali to use is lime, and this works well for hydrochloric acid; it is necessary only to pass the effluent through a bed of limestone, and a neutral solution of calcium chloride flows to waste. Calcium sulphate produced by sulphuric acid is, however, relatively insoluble, and it rapidly forms a protective coat over the lime and allows further acid to pass by unaltered. Special methods have therefore to be devised to deal with sulphuric acid. Other methods which can be employed are sedimentation and precipitation of suspended matter, and of various chemicals, and the passage of hot water through cooling towers. None of these techniques is, however, fully effective and, in all but exceptional cases, the final effluent always contains small amounts of the original pollutant.

Organic matter and many chemical effluents such as phenols, which are fairly readily oxidised, can be subjected to various processes known as biological treatments. For a popular account of these the reader is referred to Lloyd's (1947) article in *New Biology*, or for more technical accounts to Jenkins (1957) and Klein (1957b). Fundamentally two different processes are used either separately or in combination. If sewage, or some similar material, is stored in a closed container it putrefies; that is to say it uses up all the available oxygen and is then attacked by anaerobic bacteria. These continue to break down the organic matter and produce such gases as methane (CH_4) and hydrogen sulphide (H_2S). Solid matter is broken up and the whole mass becomes a foul-smelling fluid. This process may result in the loss of much of the organic matter as gas, and the liquid can often be disposed of by simple percolation into the soil. This is, in essence, the function of the rural cess-pit. It is unsatisfactory, however, for use on a large scale, except as a partial treatment. The other process is that of oxidation by means of aerobic organisms. The most widespread and easily understood of these treatments is that of filtration through trickling filters, which are the familiar structures to be seen on sewage works. The sewage is first sedimented and then passed through a bed of clinker or broken stones. The fluid passes as a trickle, so that the interstices remain full of air, and oxygen is readily available. Some months after the sewage is first turned on to a new filter, a complex growth of micro-organisms has formed on the stones, and they feed on the sewage and greatly reduce its content of organic matter. Carbohydrates, phenols, etc., are oxidised to carbon dioxide and water, proteins to carbon dioxide, water, simple amines and ammonia, and the latter may be oxidised further to nitrites or even nitrates; compounds containing sulphur produce sulphites or sulphates. In fact the organic matter becomes, in the jargon of the sewage works, 'mineralised'. The organisms which perform this work are a complex mixture of bacteria, mostly *Zoogloea*, *Sphaerotilus* and *Beggiatoa*, various fungi (Cooke, 1954; Painter, 1954) and Protozoa, particularly bell-animalcules, Vorticellidae (Rudolfs, 1941). As they feed they grow and tend to fill up the interstices and cause the filters to block up or 'pond', and if this occurs their efficiency is greatly reduced by lack of air. They are, however, normally prevented from doing this because the growths break away when they become too thick. They are also steadily eaten and disintegrated by large numbers of worms and insects which also live in the filters, where they find conditions resembling those of their natural habitat in mud flats (Lloyd, 1944). This whole aspect of biological filtration is fully discussed by Lloyd (1947), and need not be further considered here. The growth and break up of the film, however, produces suspended matter in the effluent even if none was present in the original fluid. This is a point which does concern us, as the suspended solids are light and flocculent and cannot all be removed by sedimentation. The final effluent is usually not fully mineralised, nor is it entirely free of suspended solids, and a filter is working well if 90 per cent. of the organic matter has been removed after final sedimentation.

A similar use of aerobic decomposition is the activated sludge process. In this the living organisms are grown not on stones or clinker, but as a flocculent mass, which is used to inoculate each batch of sewage. The sewage is run into tanks in which it is kept well mixed with the flocculent 'activated sludge' either by paddles or by vigorous aeration, and is thus adequately oxygenated. After a period of some hours the mixture is drawn off and settled, and the relatively clear effluent is run off. Some of the sludge is rejected, often being anaerobically digested and added, when liquefied, to the inflowing sewage, but the rest is retained to inoculate a new batch of sewage. The sludge consists of similar organisms to those present on the trickling-filter films, except that, in a satisfactory sludge, filamentous bacteria and fungi are less abundant, and Vorticellidae and other Protozoa are much more so. Even algae may be present in fair numbers and the living composition of the sludge varies with the season: the Vorticellidae are present at all times, but many other organisms appear in numbers only in the summer (Şládeček, 1958).

This massive inoculation of the sewage with actively growing organisms under conditions of good aeration performs the same functions as the filters. The general question of inoculation with suitable organisms has been discussed by Rudolfs (1941), who points out that reactions, which will occur in any event after some time, are much speeded up when a large inoculum of suitable organisms is added. For instance sewage, if stored without access to air, ultimately produces hydrogen sulphide, but only after it has built up a suitable bacterial flora. In sewers, however, where such bacteria are already present in quantity as a film on the pipes, quite a short period of stagnation leads to trouble from this gas.

Aerobic biological treatment, particularly the more easily operated trickling-filter method, has been found to be effective in the purification of various kinds of organic effluent. The filters in time build up a film of those organisms which are adapted to feed on the particular substances which are supplied to them. Thus milk wastes, phenols and textile wastes can be partially purified by biological filtration, but almost all of them give better results when they are treated together with sewage. This is probably because sewage contains adequate amounts of subsidiary substances which are needed by the bacteria, and which may be in short supply in the less varied industrial effluent. In the same way a large proportion of the toxic substances in gas-works effluent can be removed by biological treatment with sewage. The bacterium, *Thiobacillus thiocyanoxidans*, is able to oxidise the very toxic thiocyanates (Postgate, 1954), which are important constituents of gas liquor, and presumably this, or an allied organism, builds up a population in the treatment plant. There are also hopeful indications that activated sludge may be of service in the treatment of radioactive wastes (Paul, 1952), because living organisms concentrate certain elements in their bodies, which can then be removed.

However, despite the use of all these types of treatment, and of others, it is impossible to remove all undesirable properties from effluents. Hot water remains warm, poisons are still present in small amounts, sewage still contains organic matter and some suspended solids, and coal washings still contain particles in suspension.

Study of the amounts of these substances which remain is the province of the chemist, but, in all matters concerning water pollution, chemistry and biology are inextricably entangled and neither discipline can proceed far without taking account of the other.

The usual techniques used in the chemical analysis of effluents are described in several standard works (Roberts *et al.*, 1940; Suckling, 1944; American Public Health Association, 1946); and a method of determining amounts of detergent, which have recently become an important problem in sewage (Southgate, 1957a), is given in the Ministry of Housing and Local Government Report on Synthetic Detergents (1956). For a full and critical discussion of the various methods the reader is referred to Klein (1957a), Lovett (1957) and Burgess (1957). Here we are not concerned with techniques themselves, and it suffices to know that methods are available for measuring the amounts of various substances in solution, such as metals, chlorine, gases, cyanides, phenols, etc., and for the various states (reduced or oxidised) of nitrogenous and sulphur-containing ions, e.g. ammonia (NH_4^+), nitrite (NO_2^-), and nitrate (NO_3^-), and sulphide ($S^=$), sulphite ($SO_3^=$) and sulphate ($SO_4^=$). Some things are, however, very difficult to measure, amongst them being the insecticides B.H.C. and D.D.T., and, most unfortunately, oxidisable organic matter.

It is because of this last point that the Royal Commission adopted the now well-known biochemical oxygen demand test (B.O.D.). This test was originally devised by Sir Edward Frankland, who was a member of the Commission, and it is undoubtedly the most important contribution to the study of pollution which has ever been made. During the past half-century it has been much simplified and modified, and other tests, such as the amount of oxygen absorbed from potassium permanganate in a given time, are used to supplement it. It remains, however, the most important measure of the polluting power of organic effluents. although its exact significance is often not appreciated by analysts who use it. The value of these indirect tests for organic matter and their relationship to one another, and to pollution problems in general, has been most admirably discussed by Phelps (1944), and the reader who feels at home with differential equations is referred to his work for further information.

When organic matter is added to a stream it is immediately attacked by bacteria which break it down to simpler substances, and in doing so use up oxygen. The rate at which a particular type of effluent is able, in the presence of ample oxygen, to satisfy its oxygen demand depends on what it contains. Industrial effluents which contain only chemical reducing agents, such as ferrous salts or sulphides, take up oxygen by purely chemical action; they do this very rapidly, exerting what is sometimes known as immediate oxygen demand. Organic substances, such as starch, sewage and milk wastes, become oxidised only by the activities of bacteria. The rate at which they are broken down therefore depends firstly on the presence of suitable bacteria, and secondly on how satisfactory and balanced a food they are for micro-organisms. Sterile effluents, such as phenols, take some time to build up a suitable bacterial flora, and if they are very uniform in composition they may contain inadequate amounts of some substances, such as phosphates, which are needed for bacterial growth, even after they have been mixed with river water. Sewage, of course, is well inoculated with bacteria and is adequately supplied with a wide range of compounds, so it gets broken down relatively easily. But some materials, notably wood pulp, are very poor bacterial foods and are decomposed very slowly. They therefore exert a lower oxygen demand, but for a long time; and in the aggregate it may be very great. Paul (1952) quotes a figure of 1,300,000 p.p.m. for sawdust; for untreated sewage this figure is about 600.

The biochemical oxygen demand test is a purely arbitrary measure of the oxygen taken up by a sample of effluent or river water during a period of 5 days at 20° C., and it takes no account of the differences between different types of effluent. The sample is diluted with well-oxygenated water, the amount of dilution being such as to ensure that there is enough oxygen present for about 50 per cent. saturation to remain even after 5 days of incubation in a closed bottle. The initial oxygen content of the mixture is determined, and a portion of it is stored at 20° C., in the dark, to ensure that there are no complications due to photosynthesis. After 5 days the oxygen content is again determined and, from the difference between the two determinations, the B.O.D. of the original material can be calculated, making due allowance for the amount of dilution and the oxygen demand of the diluting water. This is the 5-day B.O.D., but it is not, of course, the total B.O.D. Even for sewage, 5 days is not long enough for all the oxygen demand to be satisfied. In fact sewage satisfies only about 70 per cent. of its total demand in 5 days; the 10-day B.O.D. represents about 90 per cent; and after 20 days about 99 per cent. of the ultimate demand is satisfied (Phelps, 1944). These figures demonstrate very clearly how far downstream deoxygenating effects can extend, even with so readily oxidised a mixture as sewage. This is especially apparent when one also takes into account the fact that the figures are based on experiments conducted at 20° C. This is an unusually high temperature for river water and at lower temperatures the reaction proceeds more slowly.

On similar effluents the 5-day B.O.D. test provides a useful indication of comparative 'strengths', and for sewage, which is well provided with suitable bacteria and varied bacterial foodstuffs, it is a very useful test. For other types of effluent it is less satisfactory, for a variety of reasons. The effluent may have bactericidal properties and oxidation may be inhibited or prevented except at very great dilutions. It may be almost free of the bacteria necessary to carry on the break-down of the organic matter, or it may be short of some necessary supplementary foodstuffs. For these reasons the chemist must exercise considerable ingenuity in order to obtain informative results, and I fear that this aspect of the matter is often overlooked. Many chemists treat the B.O.D. test as if it were a chemical reaction, which it is not; it is essentially bacterial. Living organisms are not merely reagents, although they can sometimes be coaxed into behaving as if they were. There have been many investigations on this test and various methods have been evolved for overcoming its particular problems. In America it is standard practice to use alkaline diluting water to make conditions optimal for bacteria, and Lea (1941) has shown that the addition of nitrates and phosphates to this diluting water leads to more satisfactory results. It is often the practice to seed the water used to dilute industrial effluents with a little sewage, thus providing suitable micro-organisms and subsidiary foodstuffs. For particularly refractory substances, such as wood fibres, which are attacked only by special bacteria, Phelps (1944) has suggested the use of river water from below the effluent outfall, where the correct bacterial flora can be expected to be present. Clearly unless such devices are used, and used intelligently, the values obtained from determinations of B.O.D. on effluents other than sewage can be very misleading: further information on the various modifications of this test is given by Klein (1957a) and Rennerfelt (1958).

Whatever its merits the B.O.D. test has one major disadvantage, which is that no results are available for 5 days after sampling. This can be a considerable handicap to a sewage-works manager who wants to know how well his plant is working. Supplementary tests are therefore used based on the amount of oxygen absorbed from potassium permanganate by effluents. The most widely used of these is a test run at 27° C. for 4 hours. This gives fairly consistent results with such complexes as sewage, and for any particular sewage the ratio of the B.O.D. to the permanganate test is fairly constant. The sewage effluents from no two towns are, however, quite alike, because they differ in, for instance, the amount of industrial wastes they contain. No general ratio can therefore be given, but usually it is about 3 or 4 for sewage, and for clean stream-waters not much above 1·0 (Lovett, 1957). For industrial wastes 4-hour permanganate-tests are less reliable because they measure only chemical oxidisability and not the availability of the contained substances as bacterial foods. By way of illustration of this fact Phelps gives the example of oxalic acid, which is rapidly oxidised by permanganate, but which, being unserviceable as a bacterial food, has no B.O.D. This difficulty can be to some extent circumvented by making two tests, one of which is a short one, usually 3 minutes, and the other the usual 4-hour test. For oxalic acid these would give the same result, but for more inert organic matter the results would differ widely.

In the United States it has become common practice to quote the polluting effects of various industrial effluents as their 'population equivalents'. This is based on their B.O.D., and although it takes no account of the differing stabilities of the various types of organic matter involved it is a useful concept, if only to bring home to manufacturers the sort of effect that their effluents are likely to have. It is known that in urban sewage each individual human contributes a total oxygen demand of about 0·25 lb. (115 g.) per day; which is equivalent to 0·17 lb. of 5-day B.O.D. Therefore a daily output of effluent which needs 1 lb. of oxygen for total oxidation has a population equivalent of 4. On this basis it can be calculated that, for example, slaughtering one cow per day equals 21 persons, tanning one hide equals 18, making 100 lb. of butter equals 34 and so on. These figures are from Phelps (1944), who gives many others, some of which are surprisingly high, e.g. manufacture of a ton of strawboard is equivalent to the daily output of 1,690 persons.

It may be asked why determination of the total amount of organic carbon would not give a more reliable picture of the actual amount of organic matter present. But apart from the technical difficulties, the way in which carbon is combined is important. For instance phenol, or carbolic acid (C_6H_6O), and glucose, or barley sugar ($C_6H_{12}O_6$), would exert quite different oxygen demands although a single molecule of each contains six carbon atoms. This will be clear from the following equations:

$$C_6H_6O + 7O_2 \rightarrow 6CO_2 + 3H_2O$$
$$C_6H_{12}O_6 + 6O_2 \rightarrow 6CO_2 + 6H_2O$$

Complete oxidation of the sugar requires only six-sevenths of the oxygen required by the phenol because it already contains more oxygen. Also, of course, many completely *biologically* oxidised substances are not completely oxidised *chemically*. Completely biologically oxidised sewage, for example, is an earthy humus-like substance, which still contains organic compounds such as lignin and which burns readily; ordinary soil is full of such substances.

Before leaving the subject of biological oxidation of organic matter we must consider the process in greater detail, and this is best done with reference to sewage, the commonest of all effluents. It contains, even after treatment, quantities of carbohydrates, fats and proteins and their breakdown products, as well as ammonia and cyanides from the gas works. As the organic substances are broken down, only about 20 per cent of their mass is synthesised into new bacterial substances. As Phelps remarks: 'This may represent a rather wasteful process of growth but it is a highly economical one in the destruction of waste organic matter.' The rest is converted into carbon dioxide and water, but at the same time nitrogen, sulphur, phosphorus and some iron and other elements are released from the proteins. At first these are in a reduced form, e.g. ammonia, amines and sulphides, and under normal conditions they do not begin to be oxidised until the oxidation of the carbonaceous matter is far advanced. The reason why this is so is not clearly understood, but it is possibly because until then conditions are unsuitable for the bacteria which carry out the oxidation. These bacteria are specialists and many of them work in production lines. For instance *Nitrosomonas* converts simple ammonia compounds to nitrite, which is then oxidised to nitrate by *Nitrobacter*. This delay in the onset of nitrification has, however, an important practical application, because the appearance of nitrates in a sewage-works effluent gives an indication that satisfactory oxidation of carbonaceous matter is being achieved. Another point about nitrates is that they can once more be reduced should oxygen be lacking at some later stage, and the same applies to sulphates, although these do not get reduced to sulphide until all the nitrate has been used up. These facts have two important consequences. Firstly, the presence of nitrate in a sample used for determination of the 5-day B.O.D., can, if the analysis is ineptly done, supply some oxygen for carbonaceous matter and so result in too low a value being obtained. Secondly, nitrate arriving in stream water acts as a sort of safety valve against nuisance, because, should the oxygen content of the water fall to zero, nitrates supply oxygen for continued break-down of organic matter. The period during which this can occur depends of course on the concentration of nitrate, but it is often long enough to prevent the further stage of reduction of sulphates and the release of evil-smelling hydrogen sulphide. Locally, out of reach of a ready supply of oxygen, these later reactions do occur, even when the general conditions are not anaerobic, which is why the undersides of stones in quite mildly polluted rivers are often blackened with ferrous sulphide. Similarly banks of sewage solids in well-oxygenated waters are always black just beneath the surface, and at greater depths entirely anaerobic processes often go on with the production of marsh-gas. This process may also of course occur with quite natural sub-aquatic soils, due to their content of plant remains, but pollutional solids can almost always be relied upon to produce bubbles when stirred. Blackening of the undersides of stones and the smell of hydrogen sulphide are particularly common as natural phenomena in the sea, as must be known to most people. Any decay in confined spaces results in de-oxygenation, but, in sea water, nitrate is in relatively short supply and sulphate is plentiful, so the reduction to sulphides occurs rapidly.

In natural fresh waters sulphate is not usually abundant and blackening and sulphides are correspondingly rarer.

The reader may have noticed that, although it takes as long as 5 days at 20° C. in an ample supply of oxygen to satisfy about 70 per cent. of the oxygen demand of raw sewage, an efficient sewage works does very much better than this and in a very much shorter time. It, in fact, reduces a 5-day B.O.D. of about 300–600 p.p.m. to something like 20 p.p.m. in a matter of hours. Part of this reduction is due to the actual physical removal of organic matter by sedimentation, but much of it is due to the biological processes in the filters or activated-sludge tanks. For example a batch of sewage under normal operating conditions passes through a trickling filter in about 15–20 minutes, yet within this brief period removal of B.O.D. to the extent of 75–80 per cent. may be accomplished, corresponding to the B.O.D. reduction achieved during about 7 days' storage with excess oxygen. At first sight it seems that the filter does not obey the ordinary laws of biological oxidation, but it can be shown that, in fact, the filter does not actually oxidise the organic matter as it received it but removes it and stores it up. The stored material is then oxidised at the normal rate and the filter holds several days' accumulation of organic matter in its slimy film. A rather similar process occurs in the activated-sludge tanks where a fresh batch of sewage is deprived of most of its organic matter within a period of an hour or two, and the sewage becomes 'clarified'. Here again the organic matter has not been destroyed, but absorbed by the sludge, and is then removed by sedimentation. The whole process of sewage treatment, and of that of similar trade effluents, is therefore essentially biological in nature, and as long as the correct organisms are present to take in and feed on the organic matter it achieves remarkable results in a very short time. But like all living things the micro-organisms that do this work are vulnerable to poisons, and as there are many tons of them in even quite small treatment works they take time to build up, and are correspondingly valuable. It is not surprising therefore that the manufacturer who tips a load of poison down a drain in the hope that it will not be noticed is unpopular with the sewage-works manager. A new filter takes months to come into full and reliable operation, and the same applies to one that has been 'killed'.

Chapter VI

PHYSICAL AND CHEMICAL EFFECTS OF EFFLUENTS ON RIVERS

WE have seen that very few effluents can be treated so as to render them completely innocuous; even the most satisfactory effluent is not river water and must therefore produce some alterations in the water course which receives it. Such alterations are ecological changes to the environments of the river flora and fauna and so are fundamental to the study of the biology of pollution.

The physical and chemical effects of pollution can, from an ecological point of view, be divided into five categories, one or more of which may be characteristic of any one effluent. Sewage, for example, is capable of producing all five types of effect. These categories are: addition of poisonous substances, addition of suspended solids, de-oxygenation, addition of non-toxic salts, and heating of the water. In practice these ecological changes rarely occur singly, but it is possible more or less to separate them for the purpose of discussing how their influence changes as one passes downstream from the effluent outfall.

Poisons usually decrease steadily in concentration. This is partly because the volume of diluting water in any river increases as more tributaries join it, and partly because many poisons, such as metals, are precipitated by chemical action, while others, particularly organic compounds, are oxidised and changed into non-toxic materials. A few poisons, of which the most important are the ammonia and sulphides produced by the breakdown of some organic effluents (Pentelow, 1953), may at first actually increase in concentration in the river. But these are exceptions; most poisons start at their maximum concentration and then steadily decline. This is shown diagramatically in *Fig.* 15.

Suspended solids also decline in amount as one proceeds downstream because they settle out of the water. The rate of settling depends on the density and size of the particles and the turbulence of the water, and is particularly rapid behind weirs and in pools or where the current is slack. If the solids are completely inert, as are mine slurries, the amount of deposit on the river bed slowly builds up and extends further and further downstream as it is stirred up and carried on by floods. Oxidisable solids on the other hand, such as those in sewage, are steadily broken down by bacteria; a balance therefore results between the rates of settling and decomposition, and the deposits slowly tail off downstream. Even so, the rate of oxidation is usually slow, and the distance affected is usually many miles. This is particularly evident where irregular distribution of the deposits results in local pockets of high oxygen demand. Thus dams and pools a long distance from the outfall may suffer from de-oxygenation although at other points the river is well oxygenated (Phelps, 1944).

De-oxygenation is usually caused by bacterial breakdown of organic matter, but it may be due to other reducing agents. In the latter event the oxygen demand is immediate; the oxygen content of the river water falls sharply and is then slowly raised by re-aeration. With organic effluents the reaction is slower, and the amount of dissolved oxygen in the river water falls more gradually. The oxygen can only be replaced by aeration at the surface or by the photosynthetic activities of green plants. As the decaying material passes downstream the oxygen deficit, which is cumulative, increases, but at the same time the rate of oxygen uptake increases because, as we have seen, it is proportional to the oxygen deficit. Thus, after some distance, as the amount of organic matter decreases, the rate of uptake of oxygen from the atmosphere overtakes that of extraction from solution. The oxygen deficit is therefore reduced, and because of the interrelationship of the two processes it follows that this occurs fairly rapidly at first and then more and more slowly as the oxygen deficit falls. This results in the so-called 'oxygen-sag curve', of which an idealised form is shown in *Fig. 16a.* If the initial oxygen demand is very great, the oxygen content of the river water may fall to zero for a long distance before re-aeration overtakes the rate of extraction from the water. In this event only anaerobic bacteria can operate in the de-oxygenated zone, and although the breakdown of organic matter continues under these conditions, and much oxygen demand may be lost to the atmosphere in the form of methane, ammonia and hydrogen sulphide, an oxygen debt, in the form of reduced substances, is built up and has to be paid off further downstream. De-oxygenation is one of the most important properties of organic matter, although it also has other undesirable attributes. For instance Pruthi (1927) and many later workers have demonstrated that toxicity is produced by septic (anaerobic) decay, and this is now known to be caused by ammonia and sulphides.

As one can measure the ultimate oxygen demand of any effluent and the rate at which this demand is satisfied, it ought, theoretically, to be possible to predict the de-oxygenating effect of a known rate of discharge of a given effluent into any particular river. This has, in fact, been done with some success on at least one occasion in America, where the calculated oxygen-sag curve in the Ohio river was found to match the observed curve very closely (Phelps, 1944). Usually, however, this is impossible, especially in small rivers, because there are so many complicating factors. The rate of oxidation is influenced by temperature, and by the presence of suitable bacteria and the necessary supplementary foodstuffs, and the rate of re-oxygenation depends on such variable factors as rate of flow, turbulence and depth of the water, and the amount of photosynthesis, all of which are ultimately controlled by the weather. Also, of course, accumulations of suspended solids exerting local oxygen-demands inevitably upset calculations.

Weirs which isolate artificial by-passes, and which allow water to pass only at times of flood, may exert devastating effects on the oxygen régime of the river below them (Butcher, 1950). The isolated reach becomes filled with water at flood-time, and this water is often more polluted than usual, because at times of heavy rainfall many sewage works are unable to handle the extra run-off from roads and roof drains and so pass quite untreated sewage out through 'storm overflows'. Normally this has little effect, because the discharge is at a time of high water and maximum dilution, but if such water is then impounded and left to stagnate it rapidly putrefies. Because of the absence of flow and turbulence its rate of oxygen uptake is low and the water becomes totally de-oxygenated; anaerobic bacteria then take over and reduce firstly nitrates to ammonia and then sulphates to sulphides. Thus a mass of water is formed which is not only de-oxygenated but which contains poisons and has a heavy oxygen debt, and when a sudden flood, such as a summer thunderstorm, occurs it is suddenly pushed into the river and passes downstream as a more or less discrete 'plug'. The sudden death of thousands of fishes, particularly in hot summer weather, can often be attributed to this sort of cause, and it may appear as if there has been a 'spill' or some other failure at a treatment plant when, in fact, everything is functioning normally.

A further complication is introduced by the presence of oil or detergents. Even a thin film of oil on the water surface reduces the rate of re-oxygenation, and a thick film may entirely inhibit it. Detergents also affect the uptake of oxygen at the water surface. These are becoming increasingly common in domestic sewage, and are posing many problems; the production of swan-like masses of foam on the rivers, which has received much publicity, is probably the least important of these. Detergents form mountains of foam at sewage works, particularly on activated-sludge plants, and these endanger the operators and get blown about causing danger to public health. They are also suspected of reducing the efficiency of the treatment process (Southgate, 1957a), and some of them are very stable and pass through sewage works unaltered. Two main types are in domestic use: alkyl sulphates and alkyl aryl sulphonates. The first, which were widely used just after the war, are the least important, as they are destroyed by biological treatment (Degens *et al.*, 1950), but unfortunately they are now less popular than the other group, which are the active ingredients of most of our present domestic washing powders. Raw sewage now normally contains about 12 p.p.m. of active detergent; about 4–6 p.p.m. passes through into the effluent (Southgate, 1957a), and it is destroyed only very slowly in river water. The result is that foaming often occurs for long distances below sewage works and, perhaps unexpectedly, it often becomes worse as one proceeds downstream. This is because, as in the washing-up bowl, the dirtier the water the less the foam. So, in the river, as the amount of organic matter is reduced by the normal process of self-purification, the persistent detergent is released to form foam, which it does very readily wherever there is broken water, particularly on damp, still days. But the effect of detergents on re-oxygenation is more important to us here than is their tendency to foam. It has been shown (Gameson *et al.*, 1955) by means of experiments on a natural stream, that when the rate of oxygen uptake is low, as in a sluggish reach, it is made even lower by the addition of detergents. There is no effect, however, when the rate of re-oxygenation is high, as in turbulent water, and it is now known that there is also none when the rate of oxygen uptake is very low indeed, as in stagnant water (Southgate, 1957a). With all these complications it will be appreciated that calculation of the probable effects of an effluent on a river, in even so apparently simple a matter as the oxygen content of the water, is possible only in exceptional circumstances.

Non-toxic salts behave in much the same way as persistent poisons in that they are steadily reduced in concentration by dilution. This applies to such salts as sodium chloride, which is a universal constituent of sewage because every packet of salt which is sold in the shops is ultimately poured down the drain. In Britain, except in some salt-mining areas, insufficient concentrations are produced in the rivers to have much effect, although the provincial visitor to London, who is struck by the rather 'flat' taste of the drinking water, is thus affected because of the high salt-content it has acquired from the sewage in the Thames. In many parts of the world, however, brines from industries and oil wells render inland waters very salt indeed. The Rhine below the Ruhr, for instance, carries no less than 15,000 tons of sodium chloride a day, and this makes its waters rather unsuitable for irrigation, with the result that Dutch farmland is now threatened by salt water from both sides (Jaag, 1955).

More universal, and more important from a biological point of view, are the so-called nutrient salts which are needed for plant growth. These are rarely discharged directly into rivers, although they may seep in from slag-heaps or be washed in when fertilisers are applied to agricultural land. They are, however, important products of the breakdown of organic matter, particularly sewage, and, as this is oxidised, nitrates, phosphates and potassium are released into the water and become available for plant growth. Below zones of heavy organic pollution the amounts of these substances are often greatly in excess of their normal concentration in natural waters, and this has important biological consequences.

Similarly the discharge of lime in any form into a soft-water river completely alters the character of the environment for living things. Hardening of the water may have far-reaching effects on the biotic community, apart from any direct effect it may have on the usefulness of the water for such industries as brewing or textile manufacture.

Heat, like poisons and salts, also declines as one proceeds downstream from the point of discharge of an effluent, partly because of dilution and partly because of loss to the atmosphere and the river bed. The rate of such heat loss depends on such purely physical factors as turbulence and wind speed, but, primarily it depends on the temperature difference between the water and its surroundings. The rate of heat loss therefore declines as the temperature falls, and because of this some difference in temperature may persist for many miles. The effect of a rise in temperature, unless it is great enough to cause sterilisation, is to speed up all biological activity and all chemical reactions. Thus hot water discharged into a river which is already organically polluted enhances the effects of de-oxygenation and, as already explained, it often brings with it its own extra load of B.O.D.

It will be seen therefore that all five of the main effects of pollution are to some extent transitory, and that if they are not too severe, and if the river is long enough and receives enough extra water from tributaries and surface run-off, it can 're-purify' itself. Indeed with the less persistent effects such as de-oxygenation, suspended solids or increase in temperature the alteration may be detectable only for a mile or two below the effluent outfall. But more usually the polluting load is heavier than this and the alteration may persist for a very long way. It has indeed been said that no river in Britain is long enough to recover completely from a load of organic matter sufficient to cause total de-oxygenation.

The reader will also appreciate that most of the effects of pollution are of the same type as 'natural' phenomena, and if they are not so severe as to produce extreme conditions they serve merely to alter one sort of river environment into another. For instance a river water which is made more silty, more acid, more alkaline, less well oxygenated, warmer, harder, saltier or richer in nutrient salts, is still a natural river water as long as the change is not so great as to overstep the bounds of normal variability. This applies even to some types of poisons. We return here to the great difficulty of defining pollution, and we must accept the fact that at least some man-made alterations to rivers closely resemble changes which in other rivers occur quite naturally.

Chapter VII

BIOLOGICAL EFFECTS OF POISONS

THE biological effects of poisonous substances are basically very simple. Generally speaking the individuals of each species of animal or plant can stand a certain amount of a particular poison, and if more is administered they die. This is an idea with which we are all familiar in its human context as it applies to therapeutic drugs and to the many poisons to which man is exposed. For most poisons one can define the threshold dose, which is the maximum amount that can be taken without causing death. Unfortunately this idea of a threshold has come to be applied in pollution studies not to *doses*, but to *concentrations*. This is a different concept altogether and has led to some confusion of thought (Wuhrman and Woker, 1958). A creature which is continuously exposed to a low concentration of poison is in quite a different predicament from one which receives a single dose, as it goes on absorbing the poison continuously. Whether in fact there is a threshold concentration below which it can survive indefinitely depends on both the creature's metabolic processes and on the nature of the poison itself. If the latter is not metabolically destroyed or eliminated it will accumulate in, or on, the creature no matter how low its concentration, and in time a fatal dose will be built up. Admittedly this may be a very long time and it may be as long as the creature's normal life span, but only in this limited sense can one speak of a threshold concentration.

On the other hand a poison which is destroyed in the body, or is steadily eliminated by some excretory process, can have a true threshold concentration depending on its rate of passage into the body and the amount that can be dealt with by the metabolic processes. Thus for some poisons it is legitimate to speak of a tolerated concentration, for others there is theoretically no concentration which is harmless; they merely become less harmful the greater their dilution. In the latter group are lead and most other heavy metals, which have been shown to damage fishes by irritating the gills and causing copious secretion of mucus which interferes with respiration (Carpenter, 1925). Here the volume of mucus secreted depends on the amount of metal which reaches the gills, so there is no theoretical concentration which will not ultimately cause death.

Nevertheless, despite these objections to the idea of a tolerated concentration of poison, it is found in practice that rivers are not devoid of life below the outfalls of poisonous effluents. Although there may be a dead zone for a longer or shorter distance, sooner or later, as one proceeds downstream to lower concentrations of the poison, creatures begin to reappear, and the further one goes into the regions of even lower concentrations the greater the number of different types of animals and plants one finds. Clearly then different creatures can tolerate different concentrations of a particular poison—which is indeed what one would expect, as both plant and animal species differ widely in their physiological processes.

An enormous amount of experimental work has been done on the effects of various poisons on fishes in attempts to find out what concentrations can be permitted to occur in rivers. This work has been reviewed by several authors (Ellis, 1937; Cole, 1941; Southgate, 1948; Huet, 1950; Doudoroff and Katz, 1950, 1953), and Jones (1957) has written an excellent discussion on the whole subject of the direct effect of pollution on fishes and the physiological reasons for it. Much of this work is of very high quality and of great theoretical interest, but for many reasons I feel that a disproportionate amount of the effort spent in all countries on the problems of pollution has been directed into this narrow field of research. Admittedly it has told us much about the actions of poisons, and their interactions between themselves and with the environment, but the final answer as to how much of any particular poison can be tolerated by *a population of fishes living a natural life in a river* still remains unanswered except where it is based on field observations. Turing (1947–9) states that 'Fish are a very useful barometer of the real state of purity of a water. No river should be considered as in a satisfactory condition unless fish will live and thrive in it'. This is indeed true, but the fishes live in the river, not in the water alone; they are the end-products of the whole complex of the environment, the plants and the invertebrates. Unless their habitat continues to supply them with the food, the shelter and the breeding sites they need they cannot thrive, even though the water may not be poisonous to them.

It used to be assumed that the lower animals are less sensitive to poisons than are fishes (Redeke, 1927), but when tests were made on the water flea *Daphnia magna* it was found to be more susceptible to many poisons than are trout (Ellis, 1937), and trout have often been shown to be more sensitive than are most species of fish (Cole, 1941; Doudoroff and Katz, 1950). More recently Grindley (1946) has shown that several species of invertebrates are killed by lower concentrations of arsenic than are minnows and rainbow trout. Undoubtedly very much depends on which invertebrates, which fishes and which poisons are under consideration. For example, midge larvae of the family Chironomidae are usually among the first creatures to appear downstream of a poisonous discharge. In the river Dove below the mouth of its copper-containing tributary, the Churnet, Pentelow and Butcher (1938) found that small green chironomid larvae were the only form of animal life. On the other hand copper salts are often used for the control of disease-carrying snails in the tropics and no damage seems to result to fisheries. To copper, then, fish are less susceptible than snails and more susceptible than some chironomids. Similarly some unpublished investigations showed that *Gammarus pulex* was about ten times less resistant to a phenolic chemical effluent than were trout, but that other invertebrates, notably worms of the family Tubificidae, were living in the small stream into which the effluent was flowing at concentrations which were rapidly fatal to trout.

One of the drawbacks of laboratory tests of toxicity is the fact that it is difficult to run experiments for long periods; thus the poisonous properties of toxins which act very slowly may be entirely overlooked. For instance Klingler (1957) has shown that sodium nitrite is definitely poisonous to minnows, but even at concentrations as high as 50 mg./l., the fishes survive for an average time of fourteen days. A short-term test made on this substance would therefore fail to detect that it was toxic except at very high concentrations. Another difficulty is the interpretation of results. It is relatively easy to try out different concentrations of poison and determine how long a particular creature will survive in them. One then finds that the survival time normally increases with the dilution at an ever-increasing rate. In theory such a relationship should give by extrapolation a concentration at which survival would be indefinite. Unfortunately, however, the relationship between survival time and concentration is often not as simple as this, and some substances actually have a maximum toxicity in an intermediate dilution (Jones, 1939a). For a full discussion of this aspect of the problem the reader is referred to Jones (1957).

Many of the earlier workers found that if one plots the concentration against the mean time for which the fish survive, one obtains a graph which is more or less a straight line over most of the lower concentrations, although at high concentrations death is so rapid that it does not occur any sooner at even higher concentrations; there is in fact a certain amount of time needed for the animal to die. The straight part of the graph, however, seemed to offer possibilities for extrapolation, because if continued it ultimately meets the ordinate at a point where the concentration could be assumed to have no effect. This was then assumed to be the threshold concentration, and many published threshold concentrations have been arrived at in this way. Herbert and Merkens (1952) showed, however, that this method is erroneous as it was found that fishes die sooner at concentrations even below this theoretical threshold than do controls in clean water. The true threshold, if there is one, must therefore be even lower than that found by extrapolation. Herbert (1952) has carried the matter further in a very illuminating discussion, which would seem to approach a final solution of the problem. He has found that for most poisons at low concentrations:

$$C^nT = K$$

where C = concentration, T = median survival time of the test animals and n and K are constants. To the mathematically minded it will be clear that, if this is so, log C plotted against log T will give a straight line which is bound to meet the abscissa (log T) somewhere. In other words the test animal is recognised as ultimately mortal, and the idea of a final threshold is dispensed with. This is probably much nearer the true state of affairs.

Wuhrman and Woker (1958) maintain, however, that, at any rate theoretically, there must be a concentration at which the animal is able to survive indefinitely, and that even at high concentrations it takes some time to show the effects of the poison. Herbert's equation is asymptotic to the axes—in other words it takes no account of these facts, and they propose that the equation should be

$$(C - C_s)^n(T - T_s) = K$$

where C_s is the minimum effective concentration and T_s, which they admit to be unimportant, is the manifestation time. This seems to me to be an esoteric refinement, and it has the disadvantage that it assumes once more that the test animal is immortal.

However, even given that it is possible by Herbert's method to arrive at a means of calculating the probable mortality that will be caused by a poison, one still has only the vaguest idea of what will happen in a river. All workers have found that fishes of one species are very variable in their susceptibility to poisons. Sometimes the susceptibility is evenly distributed about the mean, but in some experiments it has been found that, at any one concentration, it is the *logarithms* of the survival times which are thus randomly distributed (Herbert, 1952; Wuhrman and Woker, 1958). When this is so it means, in effect, that while most fishes die after more or less the same time a few live for very much longer. This raises the possibility of genetical selection of fishes which are acclimatised to higher concentrations of a particular poison than can normally be tolerated. It has indeed been found in the Upper Sacramento river in western U.S.A., which is polluted by copper, that there is a local population of fishes which seem to do well, but hatchery-reared fishes which are introduced into the river succumb to the poison (Paul, 1952). It seems probable that we have there a case of selection of the native stock for resistance to copper. Many workers have also reported that their tests have shown that fishes of different sizes, or ages, differed in their susceptibility to poisons. In some experiments small fishes have proved to be less susceptible than large ones, in others more so (Cole, 1941). We also know very little indeed about the effects of poisons on eggs and alevins, although they may prove to be more resistant to some poisons than fully developed fishes. Gardiner (1927), for instance, found that trout alevins are less susceptible to aqueous extracts of tar and phenol than are yearling fish, and that the eggs and spermatozoa are quite unaffected.

Obviously no real estimation of the effect of a poison on a *population* of fishes in a river, even discounting any indirect effects through food, can be made on the basis of tests made on only one stage of life history. One must test the most sensitive stage, and in spite of all the work that has been done we do not know this for even one poison acting on a single species of fish.

But there are other, even graver, difficulties. Effluents rarely contain only a single poison and a river may, and usually does, receive several effluents. We have already seen how chlorine acts on thiocyanates in gas-works effluent to produce a poison more virulent than either (Allen *et al.*, 1946). This is now known to be cyanogen chloride, and as can be seen from the following formula the reaction produces a considerable amount of mineral acid as well.

$$KCNS + 4Cl_2 + 4H_2O \rightarrow CNCl + KCl + H_2SO_4 + 6HCl.$$

Many other poisons have been shown to increase one another's toxicity in this sort of way (Southgate, 1932; Jones, 1939b), although whether they do so by forming more toxic compounds or by enhancing each other's physiological activity is not known. Jones (1939b) has suggested that the reduction of toxicity of copper nitrate to a variety of freshwater animals, including tadpoles, which is caused by the addition of lead nitrate may be due to some effect on the ability of the copper to enter the animals. In this instance a little lead added to a copper solution reduces its toxicity, but when more lead is added the toxicity of the mixture is increased. Undoubtedly this is a subject of considerable complexity, yet effects on the rate of penetration of poisons are probably very important. It is known that there is a fairly close relationship between the electrolytic solution pressure of the metal ions and their toxicity to fishes (Jones, 1939b); those with the lowest pressure are the most toxic. As lower solution pressure means a greater readiness to enter into combination with other substances it also implies a lower ability to penetrate, as the metal ion will more readily become combined with something on the surface of the animal. In fact the very poisonous heavy metals, lead, copper, silver, etc., kill fishes without entering their bodies at all (Carpenter, 1927), and those which do penetrate, e.g. sodium and potassium, are the least toxic (Jones, 1939b). Clearly we are still very ignorant of the detailed mechanism of toxicity, and unless we know exactly how any two poisons act singly we cannot begin to predict what effect they will have when mixed.

The toxicity of most poisons is also affected by the environment. The most important factors involved are temperature, oxygen content, pH (acidity or alkalinity) and dissolved salts, particularly calcium (Wuhrman and Woker, 1955). Because many of the earlier workers did not accurately control these variables, Doudoroff and Katz (1950) describe much of the older literature on toxicity to fishes as 'more misleading than instructive'. Nowadays, in order to obtain reproducible results, tests are made under carefully controlled conditions in fairly complex apparatus, such as that described by Herbert (1952) and Herbert and Merkens (1952), which supplies a continuous flow of standardised solution to the test-chamber.

Temperature has a direct influence on toxicity (Doudoroff and Katz, 1953) and in general, at a given concentration of poison, a rise of 10° C. halves the survival time (Carpenter, 1927; Herbert *et al.*, 1955). Poisons therefore become more poisonous in rivers during the summer.

Many substances become more toxic as the oxygen content of the water falls: such are cyanides, cresols (Herbert *et al.*, 1955; Southgate *et al.*, 1933), alkyl aryl sulphonate detergents (Herbert *et al.*, 1957), ammonia (Downing and Merkens, 1955) and doubtless many others. Also the rate of oxygen consumption of fish is altered by the presence of toxins (Jones, 1947a), and their resistance to low oxygen tensions may be impaired (Merkens and Downing, 1957).

The pH of the water has effects in its own right, and pH values of below about 5 units or much above 9 are definitely harmful to most animals. But within the normal range pH has considerable influence on some poisons. Ammonia is much more toxic in alkaline than acid water (Wuhrman and Woker, 1948; Wuhrman *et al.*, 1947) because its unionised form (NH_3) is more poisonous than the ion NH_4^+. Conversely cyanide is more poisonous in acid than in alkaline water (Wuhrman and Woker, 1948) and the same applies to sulphides, which form the more poisonous gas, sulphuretted hydrogen (H_2S), in acid waters (Doudoroff and Katz, 1950).

The dissolved salt content can also influence toxicity; in particular the presence of calcium, which was early shown to reduce the toxicity of lead (Carpenter, 1930), has since been found to do the same to other heavy metals (Jones, 1938a). Also, the content of other salts, such as sodium chloride may be important. Doudoroff and Katz (1953) state that different salts of the same metal may differ in toxicity. From this one may surmise that if a salt of a toxic metal, other than a chloride, is run into a river containing much sodium chloride it will become a mixture of the original salt and the chloride of the metal, and as a result may have an altered toxicity.

Much less work has been done on invertebrates, and many of them have proved to be awkward to work on as they are difficult to keep in the laboratory (Wuhrman and Woker, 1958; Laurent, 1958). Fish species differ in their sensitivity to poisons (Cole, 1941), and the same applies to invertebrates (Haempel, 1925; Backmann *et al.*, 1934; Stammer, 1953). For example the highest concentration of zinc tolerated by *Limnaea pereger* is 0·2 p.p.m., while water boatmen, stoneflies and caddis-worms can tolerate 500 p.p.m. (Newton, 1944). Extensive studies of tolerance have, however, been made on only a few species, e.g. the flatworm *Polycelis nigra* (Jones, 1937, 1941b), *Gammarus pulex* (Jones, 1937) and *Daphnia magna* (Ellis, 1937; Anderson, 1944, 1946) The last, a pond-inhabiting water flea, has been used as a test organism for substances likely to be discharged into rivers, on the ground that it represents a class of animals that serve as food for many fish. This is only a half-truth where stream fishes are concerned as *D. magna* is confined to ponds, but *Daphnia* has the advantage that it is cheaper and easier to work with than are fishes, It is also more sensitive than trout to many poisons (Ellis, 1937) and so should not give misleading answers.

It is not known if the various complications, such as interactions of poisons with one another and with environmental factors, which have been revealed by studies made on fish, apply also to invertebrates, but it seems probable that they do. For instance Jones (1938b) has found that the reaction between lead and copper salts discussed above applied not only to tadpoles (which are very like fishes) but also to flatworms, *Gammarus* and Tubificidae. Haempel (1925) found that different species of invertebrates were killed by widely differing concentrations of dyes used in paper manufacture. He tested a variety of snails, crustaceans and insects, and found that their order of susceptibility was different for each dyestuff, although some species were more generally sensitive than most. It has also been shown in a similar series of tests on a range of invertebrates that susceptibility to chlorine and copper sulphate is different for each species and that the order of sensitivity is not the same for the two poisons (Backmann *et al.*, 1934). Wuhrmann and Woker

(1958) have also demonstrated that the stream invertebrates *Ecdyonurus* and *Gammarus* react to poisons in the same way as do fishes and that the same mortality equation applies to them.

Several of these studies have shown also that, as with fishes, there is considerable variation between individuals. Moreover, it was found that young specimens of several species are more sensitive than adults, and there are indications that invertebrates with heavy cuticles are more resistant to chlorine than are those without. This last point suggests that, as with fishes, toxicity probably depends on the ability of the poison to enter the animal. There are therefore fairly definite reasons for believing that the general principles which apply to the effects of poisons on fishes are applicable to invertebrates also, but the available information is meagre. One is, however, able to infer a good deal about the effect of toxic compounds on invertebrates, and on plants, from data obtained from polluted rivers.

An important difference between fishes and other inhabitants of rivers lies in their mobility. Unlike plants and invertebrates they can often avoid substances which they find distasteful. They can also rapidly recolonise areas from which they have been eliminated. It was for this reason that it was early recognised (Redeke, 1927) that invertebrates show up the effect of 'spills' better than do fishes, even though it was at the time considered that the latter were more sensitive to poisons. A most striking example of the rapid recolonisation of waters by fishes was observed in California (Paul, 1952). The collapse of a retaining wall allowed a large amount of septic, and thus intensely poisonous, sewage to escape into the Tuolumne river. Hundreds of tons of fishes were killed and at one point forty tons had to be removed from a water-intake on the river. Notwithstanding this enormous mortality angling success was reported as normal within a few days of the disaster. Obviously fishes had moved in very rapidly to replace those that had been destroyed.

Fishes are repelled by some poisons, particularly acids and sulphides, but others, such as ammonia, seem to attract them (Doudoroff and Katz, 1950). In this country Jones (1947b, 1948b, 1951a) has studied the reactions of sticklebacks and minnows to various poisons, and has to a considerable extent confirmed the above statements. He has, however, shown that the concentration of the poison is important as well as its nature. For instance minnows are repelled by strong concentrations of phenols and so escape from them, but in dilute solutions the reaction is less marked and they are overcome before they succeed in getting away. The three-spined stickleback is repelled by strong solutions of ammonia but attracted by weak ones and, very unexpectedly, is attracted by strong concentrations of lead nitrate although repelled by dilute solutions. Jones has also demonstrated that, as with their killing power so with their action on the sensory perception of fishes, there can be interactions between poisons. The ten-spined stickleback avoids copper sulphate only when it is fairly concentrated, but relatively low concentrations of this salt destroy its sensitivity to alcohol, chloroform, formalin and mercuric chloride which it normally avoids. Clearly then, although we know much less about it, the subject of fishes' escape reactions is as complex as is that of toxicity.

The actual effects of poisonous substances on rivers have been investigated by a number of workers, but often not in great detail. It has many times been noted that fishes are absent for some distance below outfalls, but much less work has been done on invertebrates, and very little on plants.

In the United States Wiebe (1927) found that at one point in the upper Mississippi river tar from a gas plant had eliminated all life. In Germany it was found that where the river Pleisse contained 20 p.p.m. of phenol there was also no oxygen, presumably because of the oxygen demand of the phenol, and no life at all. But further downstream, where the phenol content was lower, various Protozoa returned, but not those species which are ordinarily associated with organic pollution (Mueller, 1954). Here presumably the controlling factor was the toxicity rather than the amount of oxygen and organic matter in the water. In South Africa, where some streams have become very acid because of drainage from mine dumps, the fauna is altered, but not in the same way as it is changed by sewage. In this area mining operations expose pyrites to the atmosphere and it becomes oxidised, producing sulphuric acid. This lowers the pH of the stream water from its normal value of about 8 to 4·5 units or sometimes even 3. Even the most extreme conditions do not eliminate all the animals, and a limited fauna persists which consists of a selection of the species normally found in unpolluted reaches (Harrison, 1958a). This selection of part of the fauna, namely those few species that can tolerate the changed conditions, is probably a fairly common result of poisonous pollution. It is for instance known that many small crustaceans can tolerate a wide range of pH as judged by their occurrence in nature, while others seem to be confined to a narrow range (Lowndes, 1952). It is therefore to be expected that the addition of acid or alkali will eliminate some species but not others.

Occasionally organic poisons exert a two-fold effect, acting first as poisons and then as organic matter when they become more diluted. In Louisiana it was found that phenolic wastes eliminated crayfish, leeches, bivalves, snails and most insects and replaced them with a typical organic-pollution community of sewage fungus, worms and *Chironomus* larvae (Lafleur, 1954). As we shall see later this 'organic-matter effect' may be more important than the toxicity of substances which might appear at first sight to be poisons.

The most thorough studies of the effects of poisons on rivers have been concerned with pollution by metals. In western Wales seepage from dumps left by lead or zinc mines has for many years given rise to complaint, and a long series of papers has been published on the subject by members of University College, Aberystwyth (Carpenter, 1924, 1925, 1926; Reese, 1937; Laurie and Jones, 1938; Jones, 1940a, b, 1941a, 1949b, 1958; Newton, 1944). Most of the mines had ceased to operate before the investigations began, and the trouble arose from the oxidation of the exposed ores (galena and zincblende) to lead and zinc sulphates which drained off into the rivers. Similar troubles have been reported from America (Paul, 1952). In west Wales the waters are soft and slightly acid, so the environment offers no protection from the metals, which are normally precipitated in hard water, and zinc has continued to leach out of some dumps for at least thirty-five years after mining operations ceased (Jones, 1958).

Carpenter's early studies (1924) showed that the affected streams contained no rooted plants, that almost the only plants which survived were the algae *Batrachospermum* and *Lemanea*, and that the fauna consisted of only a few kinds of insects. Plants such as starwort and water crowfoot, and animals such as fishes, crustaceans, worms, leeches, molluscs and flatworms, which occurred in nearby streams, were absent. It also became clear that more metal was present in the water at times of flood, when it was carried in with run-off water, than at times of low water when only springs were feeding the rivers. This was shown not only chemically but also biologically, as caged fishes were killed only during floods. Indeed in this instance the biological tests proved very significant because the analyses had been made only for lead, but the fishes died sooner in the river than would have been expected from the amount of lead present (Carpenter, 1925). This indicated the existence of another factor, which proved later to be zinc.

As the streams recovered, both in time as the exposed ore became leached out, or in space as one proceeded downstream to regions of greater dilution, a biological succession occurred. First many types of algae reappeared, together with worms and flatworms, and a few fishes at intervals. This for instance was found when metals occurred only at times of flood. Later when no more metals were present, because the superficial layers of the mine dumps had been fully oxidised and leached, starwort and water crowfoot reappeared together with fishes, molluscs (notably *Ancylastrum*) and other invertebrates. The number of animal species found in the affected reaches was 14 in 1922, 29 in 1923, 57 in 1927, 104 in 1937, when lead was no longer detectable, (Laurie and Jones, 1938), and in 1948 191 species were recorded from the river Rheidol, although this last figure includes the entire river, not only the polluted reaches (Jones, 1949b).

In the earlier stages of this work no distinction was made between the effects of lead and of zinc (Jones, 1958), but it is clear that some creatures, particularly stoneflies, mayflies and some Chironomidae, were very resistant to both, and they were found living in water containing nearly 60 p.p.m. of zinc (Jones, 1940b). Worms, leeches, crustaceans, molluscs and fishes, however, were very susceptible, as apparently were the rooted plants. Algae and caddis-worms, however, presented a peculiar problem as they were both rendered scarce by the pollution, but apparently indirectly. Investigations made by suspending slides in the river showed that the normal river algae would grow quite well in the polluted water (Reese, 1937), and it was concluded that the reason why they were scarce on the river bed was not because of the poison, but because mine debris had entered the rivers and resulted in an unstable shifting bed. Jones (1949b, 1958) has suggested that this may also account, at least in part, for the absence of rooted plants. The absence of caddis-worms appeared to be a direct result of the shortage of the algae on which they feed. They were in fact found occasionally in the polluted water (Jones, 1940a, 1940b) and can tolerate high concentrations of zinc (Newton, 1944), but those species which eat algae were scarce where the river bed was covered with loose mine-grit. Carnivorous caddis-worms, however, such as *Rhyacophila* and *Polycentropus*, were common in this area, and some detritus feeders such as the stonefly *Leuctra* were particularly abundant there, possibly because of lack of competitors (Jones, 1940a). Jones (1958) has in fact suggested that the lack of variety in the fauna of the zinc-polluted river Ystwyth may be due to shortage of algae and the consequent lack of variety of food.

The effect then of pollution with lead and zinc is simply the elimination of some of the fauna and flora, although the observed effects on the latter may have been physical rather than chemical, and, as found in streams polluted by mineral acid, the survivors are normal river creatures. No special 'pollution fauna' develops, although the surviving species may be more abundant.

An instance of pollution by copper has been studied in the river Churnet, a tributary of the river Dove (Pentelow and Butcher, 1938; Butcher, 1946b, 1955). This river was, at the time the work was done, polluted by organic matter, but was showing signs of recovery a few miles further downstream (see Chapter IX). It then received a copper-works effluent which raised the copper content of the water to 1 p.p.m. and sometimes more. This was found to have a very striking effect on the flora and fauna. All the animals, which above the works consisted of Tubificidae, *Chironomus*, *Asellus*, leeches and molluscs, disappeared. None was found in the next 11 river miles to the confluence with the Dove, where the copper content had fallen to 0·6 p.p.m. The algae were also seriously affected; above the works they consisted of the normal species found in water recovering from pollution, *Stigeoclonium*, *Nitzschia palea*, *Gomphonema parvulum*, *Chamaesiphon* and *Cocconeis*, and about 1,000/sq. mm. grew on test slides which were suspended in the water for three weeks. Below the outfall this number fell sharply to 150–200/sq. mm., and two unusual species, the small spherical *Chlorococcum* and the diatom *Achnanthes affinis*, made up most of the flora. This condition persisted for at least 3 miles, but 5 miles below the outfall the numbers had risen steeply to over 33,000/sq. mm., and further downstream, and into the Dove over 30 miles below the outfall, numbers of over 50,000/sq. mm. were recorded. These enormous numbers were made up not only of the two unusual algae, but of the original algae also, although *Cocconeis* did not reappear for 30 miles.

Clearly then algae, like animals, are differently affected by poisons and some reappear sooner than others. The enormous numbers which developed were probably due to the absence of animals to graze them. Even in the Dove where, below the mouth of the Churnet, the copper content was only 0·12 p.p.m., animals were scarce and consisted only of a few green chironomid larvae. Above the Churnet the investigators found 478 animals/tenth of a square metre and 30 different species. Below the confluence these figures were 17 and 2, and even 20 miles farther on they were only 108 and 10, and although most groups of animals were represented molluscs and shrimps were still absent. Even a very low concentration of copper can therefore produce devastating effects. In contrast, however, to the effects observed on the animals and algae, the weeds in the Dove seemed to be unaffected, and were abundant and healthy even where the copper content was 0·12 p.p.m.

From the information given above we can draw general conclusions about the effect of toxic pollution, which are illustrated diagrammatically in *Fig.* 15. This shows that after a poison enters the river its concentration slowly declines, either because of dilution, precipitation or destruction. The animal species are eliminated and then slowly reappear in small numbers. The algae are at first reduced, and the species that survive as dominants may be unusual. They may then build up to large numbers, because of the absence of animals to eat them, and then slowly decline. Under special circumstances, however, as in the Welsh rivers, this algal increase may not occur, and there may, of course, be poisons which affect the algae as much as, or more than, the animals. It is, for instance, known that blanket-weed, *Cladophora glomerata*, is sensitive to iron salts (Blum, 1957), and there is no evidence that this applies to most animals. The data on higher plants are too conflicting for them to be included in the diagram.

Thus far we have been considering cases of fairly serious pollution, where the effect is very striking. Biological investigation can, however, show the effects of very mild pollution. Table 8 shows some of the results of three surveys made on a river which was receiving an effluent containing, on occasion, very small amounts of ammonia and cyanides.

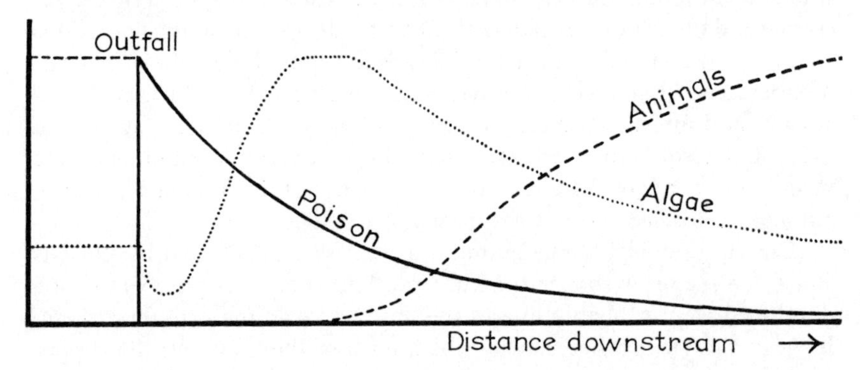

FIG. 15. Diagrammatic presentation of the decrease in concentration of a poison in a river and the corresponding changes in numbers of algae and numbers of species of animals.

The figures shown in the table are the numbers caught with an ordinary hand-net during a fixed time, and only selected organisms are shown. Others have been left out because their occurrence was too spasmodic to illustrate anything, or because they were scarce. The three surveys were made in the Aprils of succeeding years and so are strictly comparable. The maximum effect of the effluent was, of course, found at station D, where it was still flowing along the bank of the river. Station C, on the opposite bank, was, although below the effluent outfall, still effectively above it, but mixing was complete at station E. Examination of the table reveals several things. Firstly some organisms, *viz. Polycelis*, *Esolus*, *Leuctra*, *Isoperla* and the alga *Lemanea* were considerably affected by the effluent in the immediate vicinity of the outfall, but all of them, except *Leuctra* (and *Isoperla* in 1957), built up their numbers again within a very short distance. Some animals, among which were the mayflies, *Amphinemura* and Orthocladiine Chironomidae, were less affected, but their num-

TABLE 8

Numbers of animals collected by a standardised netting technique, and estimates of the abundance of plants, at various points in a river very slightly polluted by a poisonous effluent. Only selected organisms are shown. There are three sets of samples collected in the Aprils of 1955, 1956, *and* 1957, *and these are distinguished by different types (ordinary* = 1955, *bold* = 1956, *italics* = 1957). (*P* = *present,* C = *common,* A = *abundant). Although station C was below the outfall it was outside the course of the effluent, which flowed down the right bank and exerted its maximum effect at station D. Mixing was complete at station E.*

Station No.	A	B	C	D	E	F	G	H	I	J
Distance above outfall yd.	2,200	50	—	—	—	—	—	—	—	—
Distance below outfall yd.	—	—	100 left bank	100 right bank	600	700	950	1,200	1,600	1,900
Flatworm *Polycelis felina*	5	1	19	6	24	123	38	21	26	7
	1	**2**	**8**	—	**1**	**4**	**1**	—	**1**	**5**
	1	*2*	—	—	*3*	*2*	*13*	*1*	*1*	*4*
Mayflies *Baetis rhodani*	56	44	67	20	131	56	28	25	32	20
	61	**18**	**36**	**2**	**6**	—	**1**	—	**2**	—
	329	*347*	*475*	*9*	*43*	*40*	*51*	*42*	*63*	*58*
Rhithrogena semicolorata	267	263	214	31	375	62	53	119	54	47
	149	**88**	**253**	**2**	**6**	**3**	**11**	**14**	**2**	**54**
	102	*169*	*239*	*4*	*47*	*8*	*2*	*8*	*2*	*2*
Stoneflies *Amphinemura sulcicollis*	62	119	60	23	48	45	13	24	39	27
	228	**175**	**166**	**11**	**34**	**12**	**48**	**15**	**23**	**70**
	130	*267*	*185*	*42*	*131*	*24*	*38*	*19*	*15*	*8*
Leuctra spp.	22	27	16	—	2	1	—	3	3	2
	4	**6**	**8**	—	—	—	—	—	**2**	—
	89	*57*	*18*	*1*	*4*	—	*1*	*1*	—	—
Isoperla grammatica	17	30	10	5	14	8	1	3	1	1
	69	**105**	**132**	**23**	**13**	**3**	**18**	**6**	**6**	**8**
	43	*41*	*29*	*2*	*10*	—	—	—	—	—
Caddis-flies *Rhyacophila dorsalis*	4	5	2	3	5	9	3	3	5	5
	12	**6**	**5**	**1**	**2**	**1**		—	—	**3**
	13	*19*	*8*	*10*	*7*	*7*	*5*	*9*	*15*	*10*
Hydropsyche instabilis	—	—	—	—	—	—	—	1	—	—
	3	**9**	**6**	**1**	**9**	**9**	**7**	**1**	**8**	**11**
	39	*55*	*22*	*19*	*27*	*15*	*40*	*21*	*40*	*36*
Sericostoma personatum	3	4	1	1	1	12	2	5	6	1
	1	**4**	**14**	**1**	**5**	**3**	**3**	—	**1**	**5**
	3	*2*	*4*	*1*	*1*	*1*	*4*	—	*2*	*3*
Beetle (Helmidae) *Esolus parallelopipedus*	53	94	100	2	23	13	2	17	8	23
	11	**6**	**38**	—	**4**	**4**	**2**	**2**	**7**	**7**
	40	*55*	*12*	—	*3*	*11*	*14*	*11*	*10*	*12*
Biting midges Ceratopogonidae	4	—	—	2	1	3	—	—	—	—
	26	**11**	**38**	**14**	**10**	**14**	**12**	**12**	**21**	**18**
	2	*13*	*3*	*23*	*41*	*20*	*21*	*13*	*15*	*10*
Non-biting midges Orthocladiinae	109	76	153	66	51	38	25	104	20	85
	70	**163**	**100**	**37**	**40**	**69**	**90**	**103**	**10**	**44**
	37	*48*	*25*	—	*23*	*23*	*38*	*9*	*10*	*7*
Tanypodinae	—	—	1	1	—	1	1	1	1	1
	—	**1**	**3**	**1**	**2**	**4**		**1**	**6**	**2**
	2	*1*	*1*	*6*	*7*	*9*	*6*	*6*	*2*	*9*
Alga *Lemanea*	C	C	C	P	P	C	C	C	C	C
	A	**A**	**A**	—	**P**	**P**	**C**	**C**	**A**	**C**
	A	*A*	*A*	—	*P*	*C*	*C*	*A*	*A*	*A*
River moss *Hypnum palustre*	C	P	P	C	C	A	A	A	A	A
	P	**C**	**C**	**C**	**C**	**A**	**A**	**A**	**A**	**A**
	C	*P*	*C*	*P*	*C*	*A*	*A*	*A*	*A*	*A*
Total no. of animals in sample	654	802	698	288	278	452	223	405	285	325
	680	**653**	**931**	**136**	**187**	**246**	**230**	**164**	**133**	**304**
	976	*1,158*	*1,105*	*336*	*394*	*187*	*255*	*154*	*191*	*184*

bers were slightly reduced. Others, such as the caddis-worms, the midges Ceratopogonidae and Tanypodinae and the moss *Hypnum*, were quite unaffected.

This differential effect on different members of the stream community is, as we have seen, characteristic of poisonous pollution, but here we have an illustration of the process in a very mild form. Fishes in this instance were quite unaffected, as they remained in large numbers in a pool just below station E and were frequently seen at other stations, including even D. Another point which the table illustrates is the steady falling off in numbers of some creatures, and of the total number of animals caught as one proceeds downstream. The major drop, at station D, was, of course, due to the effluent, but numbers were building up again at station E, and the subsequent drop in numbers had nothing to do with the effluent and was due to a change in the character of the river bed, which is less gravelly, more rocky and also more silty below this point. A third point shown by the table is the difference between years. In 1957 *Baetis* was particularly abundant and so swelled the totals collected at stations A, B and C; in 1956 *Isoperla* was rather more abundant and in 1955 biting midges and *Hydropsyche* were scarcer than usual. These are merely normal biological phenomena. In 1955 the effluent was exerting very little effect except at station D, but in 1956 the effect was more severe and some creatures became relatively scarcer below the outfall or were eliminated for some distance: e.g. *Rhithrogena* and *Leuctra*.

In 1957 there were clear signs of recovery. The only creatures not showing this were *Isoperla* and perhaps Orthocladiinae; both were however, rather less abundant everywhere than they had been in 1956 and this has probably obscured the picture a little. The figures in the table can be correlated with changes in the effluent, which deteriorated slightly during 1955. New treatment plant was installed, however, during the summer of 1956, and when it came into operation the river was restored to its original condition. The effects of the effluent were then very slight and extremely localised, but were nevertheless biologically detectable.

This example therefore illustrates how sensitive a measure of stream conditions biological examination can be, and that even where an effluent has no effect on fishes, and practically none on water chemistry, its action can be seen in the river. In this instance the stonefly *Leuctra* and the alga *Lemanea* demonstrated the presence of the effluent very clearly, but during the period under review the river water only rarely contained detectable amounts of ammonia and cyanides. Thus the chances are that chemical analysis would have shown nothing or, had samples been collected on one of the few days when fair quantities of these poisons were present, a very misleading impression would have resulted. In fact this was a very marginal case of pollution, quite unexceptionable from a fisherman's point of view but nevertheless detectable biologically.

Chapter VIII

BIOLOGICAL EFFECTS OF SIMPLE DE-OXYGENATION AND SUSPENDED SOLIDS

THE biological effects of chemicals, such as sulphites, which merely de-oxygenate the water have, so far as I am aware, received little study. In practice most of these substances rarely occur alone; they are usually only components of other, usually organic, effluents, the effects of which they enhance. This applies, for instance, to sulphite liquor from wood-pulp mills, which has been reported to eliminate mayflies and molluscs from long stretches of rivers in Louisiana (Lafleur, 1954). One type of chemical de-oxygenator does, however, often occur alone, namely the ferrous salts, bicarbonate or sulphate. These compounds are frequently present in underground waters, and some springs and many mine adits contain fairly large amounts, with the result that the streams to which they give rise are at first completely free of oxygen and quite lifeless. Such waters can always be recognised because, as one follows them downstream, the bed becomes rusty-red with the ochre (ferric hydroxide) resulting from the oxidation of the ferrous salts. Thus, as the oxygen demand becomes satisfied and the water oxygenated, a new type of pollution is produced; that of suspended solids.

Microscopic examination of the iron deposits shows that they usually consist very largely of filamentous bacteria. These are sheath-bacteria, in which the individual cells lie end to end in a gelatinous coat in which the iron is deposited: the commonest species is *Leptothrix ochracea*. It is suggested that, in iron-containing waters, the bacteria obtain the necessary energy for the manufacture of organic matter from carbon dioxide, salts and water, by oxidation of the ferrous salt. They seem, however, to be unable to live entirely without organic matter, and although they can tolerate very low oxygen concentrations they are unable to live without any at all. In a mine adit in North Wales, which I have had under observation for many years, they are absent from the deposits near to the source, where there is never any oxygen, but become commoner further downstream.

Conditions similar to those in mine-drainage waters are sometimes produced by waters seeping from swamps. The reader will be familiar with the iron-stained runnels and ditches which are often to be seen in waterlogged areas. Usually the rate of flow of such iron-containing waters is small, but where large volumes enter a river the effect is first to eliminate all those creatures which cannot stand a low oxygen content—that is to say the inhabitants of eroding substrata—and then to produce the symptoms of pollution by inert suspended solids.

The effects of inert solids are twofold. If they are light or very finely divided, as are some mine slurries or the waste water from china-clay quarries (Pentelow, 1949), they do not settle readily but they make the river water opaque to light and so render all plant and algal growth impossible. There seems to be little evidence of any direct effect of suspended matter on animals, although in Denmark fish have been reported as having been suffocated by ochre produced by the drainage from a lignite mine (Larsen and Olsen, 1950); and Paul (1952) reports that quartz grit from stamping mills has damaged fishes in California by abrasion. Generally, however, fishes are only harmed by concentrations of suspended solids which could only occur in a river under very exceptional conditions (Cole, 1941; Van Oosten, 1945; Wallen, 1951), and Pentelow (1949) has reported that sea trout pass regularly up the river Fal, through the reaches which are badly polluted by china-clay, to spawn in the clean headwaters. The effect on spawning sites is dealt with below. He also found a few ephemeropteran nymphs in Torry brook a tributary of the Plym, the waters of which were milk-white with china-clay. In America it has been suggested that turbidity and the consequent absence of plants are the primary consequence of pollution with inert solids, and that animals are only secondarily eliminated. Surber (1953) found that ochre produced by mine effluents in the Menominee river, Michigan, eliminated plants and reduced the fauna from 24 species to 2. The reduction in number of animal species was, however, probably due at least in part to the smothering effect discussed below.

Another effect of very fine suspensions is that, as argued by Pentelow, even if they do not eliminate all life from a river they must make it difficult for fishes to feed. Most fish hunt by sight and they are presumably hampered by a fog of suspended matter. Perhaps one finds support for this suggestion in the well-known poacher's trick of fishing for trout during spates with a large worm. Flood water is usually silty, giving poor visibility, and a large pale-coloured worm is more likely to be seen than most of the trout's normal food animals.

Suspensions of very fine solids can therefore eliminate all the life from a river, or reduce its amount without greatly altering its composition, simply by shading out all or some of the plant life. They do not usually affect migratory fish which have only to pass through a polluted reach, but they may damage other fishes directly. Perhaps the most notable instance of this kind was the extermination of the char of Ullswater. These fishes used to spawn in only one of the streams flowing into the lake which became the recipient of a mine slurry (Tate Regan, 1911). No material was being deposited on the stream bed, at any rate in 1939 some years after extermination was complete, but the water was quite opaque and no invertebrates were present. Exactly why the char failed to continue to breed in this stream is not known, but it may well have been the complete absence of food for the newly hatched fry. It is possible that some toxic effect was involved, but there were no complaints of dead fishes in the lake near the stream mouth, so this seems unlikely.

The second effect of inert solids occurs when they settle out of the water on to the stream bed, which happens when the particles are large or heavy or when the current is slack. The deposits smother all algal growth, kill rooted plants and mosses, and alter the nature of the substratum. We have already seen that quite large-sized mine-grit can, by rendering a river bed unstable, lead to a decline in the number of weed beds (Jones, 1943). Quantities of silt-like material destroy plants even more effectively, and quite small amounts, such as those produced by soil erosion or the regular washing of farm implements or root crops, may change the nature of the stream bed sufficiently to alter the flora. For example the non-silted community of macrophytes (Chapter III) may be replaced by the silted community, plants like *Potamogeton pectinatus* replacing *Ranunculus* and *Myriophyllum* (Butcher, 1933).

The primary effect of the settling out of solids is to destroy, or in mild cases to alter, the vegetation, and this produces a corresponding change in the fauna. But there are also direct effects on animals. Sediments falling onto eroding substrata fill up the interstices between the stones, thus depriving cryptic animals of their hiding-places. At times of slow flow they also coat over the stones and so render ineffective the various hold-fast mechanisms of the stone-fauna, all of which depend on the presence of a smooth firm surface. It is found therefore that all or most of the typical fauna disappears and is replaced by burrowing or tube-building creatures, such as worms and chironomid larvae, and the numbers of these which appear depend on the availability of food.

An example of this effect is shown by a small tributary of Ditton brook, near Liverpool, which passes close to a colliery and is used for coal-washing. On leaving the colliery grounds, the stream water is black and quite opaque, and the stream bed is completely covered over with several inches of black muddy grit together with leaves and twigs from nearby trees. Despite its appalling condition there is a dense fauna consisting of chironomid larvae of several types, including the red larvae of *Chironomus*, and worms of the families Tubificidae and Naididae. These are all inhabitants of soft substrata, and very different from the normal fauna of streams in the neighbourhood. The unpolluted headwaters of Ditton brook are stony, with frequent patches of *Ranunculus* and a fauna consisting largely of *Gammarus* and the mayfly *Baetis rhodani*, together with caddis-worms, flatworms, leeches and snails; although some chironomid larvae are present they are not important constituents of the fauna, and the same applies to the Naididae and Tubificidae. Here then the pollution has selected part of the fauna and allowed it to build up to large numbers, but the normal animals of eroding substrata have gone.

As so often happens, however, when one tries to find a simple example to illustrate a point, there are certain features about the fauna of this stream which suggest that the pollution it receives is not entirely due to inert dust from coal-washings. These are the unexpected abundance of the animals and the presence of large numbers of *Chironomus thummi*, which is, as we shall see, a species which is particularly abundant in organically polluted water. This led me to suspect the presence of organic matter as well as coal-dust, and further investigation revealed a large pig-farm upstream of the colliery. This example has been included to illustrate how quite a simple and brief biological examination may indicate the presence of types of pollution which are not suspected. A chemist examining this stream below the colliery would certainly measure the suspended solids, although the black opacity of the water might make him feel that even this was hardly necessary.

Should he also measure the B.O.D. he might be suspicious when he found it to be high, but he would perhaps explain it on the ground that the stream bed was full of debris from the trees. A few minutes spent studying the fauna, however, showed that chemical analysis should include the tests normally made for organic pollution and that it was worth searching for sources of sewage or other organic pollution. This can, of course, be done both chemically and biologically, but the latter method is much the most rapid. It is merely necessary to follow the biological signs of organic pollution upstream to their point of origin (see Chapter IX), and this, in a small stream, takes little longer than it does to walk its length.

The direct effects of deposits of inert solids on depositing substrata are less marked, but even here all but the burrowing animals, primarily Tubificidae and *Chironomus*, find conditions intolerable because they are continually smothered. In America, for instance, the silt from soil erosion has been observed to destroy mussels (Welch, 1935). When the effect was mild only young specimens were overcome, presumably because being smaller they could not move upwards as fast as mature specimens, but where the load of silt was heavy older specimens were also killed. In Norway large areas of stream bed have been blanketed out in this way by sawdust (Redeke, 1927).

Sawdust, wood fibre and some fibrous textile wastes are not totally inert and, being organic substances, are capable of slow bacterial breakdown. The carpet of deposit therefore decays and produces de-oxygenation and sewage fungus (see Chapter IX). Simmonds (1952) reports that in Spanish River, U.S.A., wood-fibre deposits became matted with fungus and a few algae; they then entrapped gases, presumably methane caused by anaerobic decomposition in their lower layers, and floated off, extending their influence even further downstream. Where the deposits occurred he found only worms. Similarly in the river Holme, Yorks, members of the West Riding Rivers Board (Garner *et al.*, 1936) found that sludge-like deposits of textile materials, chiefly wool and cotton fibres, produced fungus and local de-oxygenation. In the upper unpolluted reaches the river contained a rich and varied fauna, but where it was affected by the deposits there were only flatworms, worms, leeches, chironomid larvae, snails and limpets. The presence of flatworms and limpets indicates that in this instance the pollution was not particularly severe, but even so sponges, shrimps, mites, stoneflies, mayflies, caddisworms, beetles and *Simulium*, which were all present upstream, had been eliminated.

Suspended solids in even quite small quantities may have a serious effect on the spawning sites of salmon and trout. These fishes bury their eggs in gravel (Jones J. W. and Ball, 1954), and they are very selective of the particular area which they use for spawning. It has been found that the places where they spawn are always those where water flows through the gravel (Stuart, 1953; Jones J.W., 1959), such places occurring, for example, where a riffle forms the lower boundary of a pool. The moving water carries oxygen to the eggs, but should the interstices in the gravel become blocked by silt the eggs are asphyxiated (Jones J. W., 1958). Damage of this kind has been reported in western America as the result of silt from mines (Paul, 1952). Clearly quite small amounts of suspended solids may cause this sort of trouble, as silt tends to be deposited in just those areas which are most suitable for egg-laying.

In conclusion then we may say that suspended solids tend to eliminate algae and plants, and that they alter the fauna by blanketing over the stream bed. In extreme cases only worms (usually but not always Tubificidae) and Chironomidae survive, and the numbers of these depend on how much food is available for them. In milder cases the normal stream-fauna is only reduced in quantity and may not be much altered in quality, although there is evidence that stoneflies, mayflies and caddis-worms are more affected than some other creatures. Probably further studies of this type of pollution would indicate finer distinctions than are at present known. For instance, one would expect the net-spinning caddis-worms to be affected before the case-building types, simply because of the clogging of their nets, and that *Amphinemura sulcicollis*, which tolerates rather silty microhabitats in unpolluted stony streams, would be the last of the stonefly species to succumb. Similarly, among the mayflies, *Caenis* and *Ephemera* tolerate silt whereas the flattened nymphs of the Ecdyonuridae do not. These are, however, speculations which are not based on observations on rivers polluted solely by inert suspended solids.

Chapter IX

BIOLOGICAL EFFECTS OF ORGANIC MATTER

POLLUTION with organic matter is very complex, as it involves not only the de-oxygenation of the water, but also the addition of suspended solids, the organic matter itself, and poisons such as ammonia and sulphides and in sewage one often finds also cyanides. It is therefore not possible to discuss the biological effects of each aspect separately, as has been done for poisons, simple de-oxygenators and inert suspended solids; indeed in our present state of knowledge we can only partially disentangle the various factors from one another.

There have been many biological studies of rivers polluted by a wide variety of organic effluents, and they have shown that all such substances produce broadly similar results. Moreover the general picture has been found to be the same in Europe (Redeke, 1927) as it is in North America (Wiebe, 1927; Campbell, 1939; Gaufin and Tarzwell, 1952), and the same doubtless applies to other parts of the world. Despite the general similarity of the effects of organic effluents there are, however, differences in detail between those of different types. Thus, for instance, it was found in Denmark that drainage from silage induced a heavy growth of sewage fungus without causing much de-oxygenation (Rasmussen, 1955), and the same phenomenon was observed below a kraft-pulp mill on a river in Idaho (Wilson, 1953). These two effluents, although similar in their biological action on the rivers, are very different chemically: silage juice is very acid and has no offensive smell, whereas kraft-pulping of wood produces an alkaline effluent which is very evil-smelling because of the presence of mercaptans (Van Horn, 1949). A similar production of large amounts of sewage fungus without much de-oxygenation was observed in this country, where the Bristol Avon was polluted by milk wastes (Pentelow *et al.*, 1938), and it is a common result of pollution with wastes from wood-pulping mills (Stundl, 1958). In broad outline, however, the basic biological effects of all organic effluents are similar, and this applies unexpectedly even to such materials as phenolic wastes, which both in this country and in North America (Lafleur, 1954) have produced effects very similar to those of sewage pollution.

Fig. 16, making use of an idea introduced by Bartsch (1948), demonstrates these general effects diagrammatically. In sections A and B the purely physical and chemical phenomena, and the way in which they change as one passes downstream, are shown. These have already been discussed in Chapter VI. The example shown is a case of fairly severe pollution, but not so severe as to cause complete de-oxygenation. Section C shows the corresponding changes in micro-organisms. The bacteria are at first abundant and then decline as the organic matter is used up; sewage fungus appears, increases, and then declines in the same way. When oxygen conditions permit, large numbers of Protozoa are built up and most of them belong to species which feed on the bacteria. The algae at first decrease in numbers; but then, as nutrient salts are released from the organic matter and conditions improve, they increase greatly in numbers, and then decline as the supply of nutrient salts is diminished. Blanket-weed, *Cladophora* (*Fig.* 4), appears very abundantly as recovery begins, and then declines like the other algae. Section D shows the effects on the larger animals. Fishes and clean-water invertebrates decline rapidly in numbers and usually all disappear as they are overcome by lack of oxygen, the blanketing effects of suspended solids and sewage fungus, or release of poisons. They then reappear in small numbers and little variety, and their numbers and variety increase steadily as recovery takes place. In the badly polluted zone the clean-water animals are replaced by a very abundant 'pollution fauna' consisting largely of sludge worms (Tubificidae), blood-worms (*Chironomus*) and the water louse (*Asellus*). These three types of animal succeed one another in importance as one proceeds downstream from the outfall. A point that should be stressed here is that although it is legitimate in this context to speak of a pollution fauna, none of the animals which belong to it are confined to polluted water. All are indeed, as we already know, normal inhabitants of river muds and ponds, and they merely happen to be favoured by organic pollution.

This then is the general picture. Let us now consider each of these biological phenomena in turn, without losing sight of the fact that they are interrelated and that they succeed one another in importance in the way indicated in the diagram.

Bacteria

Untreated sewage already contains large numbers of bacteria, some of which are of faecal origin such as the well-known *Escherischia* (*Bacterium*) *coli*, and many disease organisms. Many of these are lost during biological treatment (Southgate, 1951) but others survive, and large numbers of saprophytic bacteria are acquired from the sewage works. In the same way an industrial effluent which has been biologically treated is well

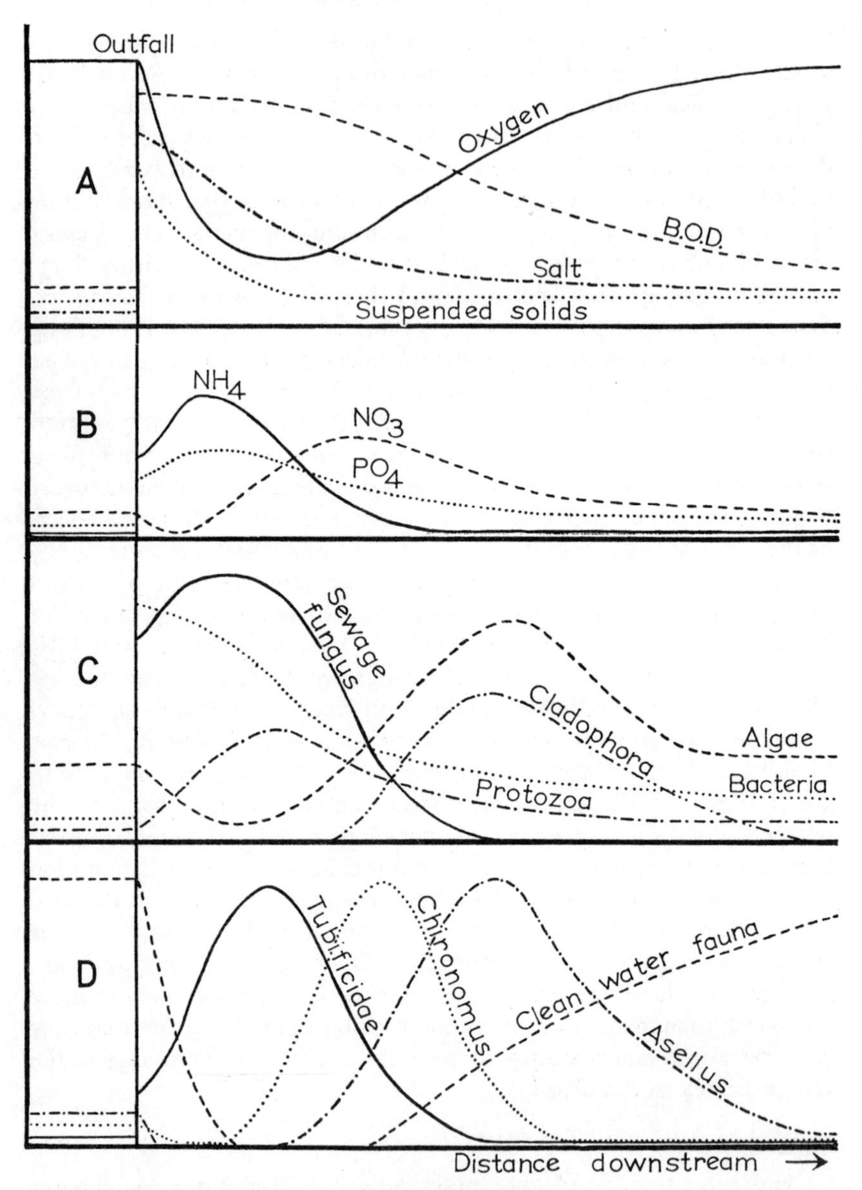

FIG. 16. Diagrammatic presentation of the effects of an organic effluent on a river and the changes as one passes downstream from the outfall. A & B physical and chemical changes, C Changes in micro-organisms, D Changes in larger animals.

inoculated with those species of bacteria which are able to carry on the breakdown of the particular type of organic matter which it contains. Some effluents are sterile, but after they have been discharged continuously into a river for some time a population of suitable micro-organisms is built up in the water. If, however, the effluent has bactericidal properties, the whole biological succession may be inhibited and no change in the composition of the organic matter may occur for many miles (Southgate, 1951). This is of course toxic pollution, the biological effects of which have already been discussed. Ultimately, however, dilution or destruction of the toxin allows bacteria to develop, and then the normal biological succession associated with the breakdown of organic matter occurs.

All organic effluents, therefore, ultimately become inoculated with suitable bacteria, the numbers of which may be very high near the outfall, and decline downstream only when the concentration of suitable foodstuffs has been considerably reduced. The range of types of bacteria involved is very great. and many of them are specialists. We have already seen that nitrification is carried out by certain genera; in the same way particular types break down cellulose, starches, fats and proteins, and others attack their breakdown products. There are also several species which inhabit totally de-oxygenated water, or the anaerobic layer of the mud in places where the overlying water contains some oxygen, and produce methane from simple organic compounds (Hawkes, 1957); hence the production of bubbles of gas from deposits of sewage solids. Others reduce sulphates to sulphides under anaerobic conditions, thus releasing the obnoxious smell of hydrogen sulphide, and blackening mud and sand where this gas combines with iron to produce black ferrous sulphide. These bacteria include the small comma-shaped *Desulphovibrio desulphuricans* (Postgate, 1954).

Most bacteria can be detected only by special culturing techniques, but a few can be recognised under the microscope. Among these are the large active spiral-shaped *Spirillum* species, which are abundant in waters heavily polluted by sewage, and several species which reduce sulphates. Many of these are described and figured by Liebmann (1951), but a good microscope is necessary to see them satisfactorily. Methods for the detection and estimation of the numbers of other types of bacteria, including the most important pathogenic species which occur in water, are given by Olszewski and Spitta (1931) and Suckling (1944).

Sewage fungus

A few species of micro-organisms form massive colonies in organically polluted water. These can be seen readily with the naked eye and are collectively referred to as 'sewage fungus' (*Fig.* 17, *Plates* 1 and 2). Only some of the organisms involved are, however, strictly speaking fungi; some are colonial bacteria and others are animals. This type of growth develops anywhere where there is a suitable supply of nutrients, and we are all familiar with it as the slimy growths in sink outflows and on drain covers. It has been the subject of a number of special studies, and the various organisms have been described by several workers (Butcher, 1932a; Liebmann, 1951; Cooke, 1954). It forms ragged white, yellow, pink or brown masses on all solid objects in the river, and may even form a carpet over mud surfaces. At times of high water it breaks away and drifts downstream (Butcher *et al.*, 1931) and the drifting masses may continue to grow in the open water of large rivers, forming a sort of plankton (Liebmann, 1953). Drifting pieces of sewage fungus cause trouble to fishermen, as they clog lines and nets, and they also carry the effects of the pollution downstream. Once they have left the zone where active growth is possible they become moribund and later decay and, particularly in pools where they collect on the bottom, they form local pockets of de-oxygenation. Sewage fungus is also one of the most unsightly products of pollution.

The most important constituent is the sheath-bacterium *Sphaerotilus natans*, in which the individual cylindrical bacterial cells are embedded end to end in a slimy gelatinous filament (*Fig.* 17). The growths have various forms, the filaments adhering to one another in different ways or the cells differing slightly in appearance. One form appears to branch and has received the name *Cladothrix dichotoma*, but several investigators (Butcher, 1932a; Lackey and Wattie, 1940; Pringsheim, 1949; Bahr 1953) have shown that all these different types are growth-forms of a single species. Moreover Pringsheim (1949) has suggested that the common iron bacterium *Leptothrix ochracea* is also a form of *S. natans*. In the commonest form the filaments are free and unbranched; aggregations of filaments occur rather rarely, and only when conditions for growth are not good do branched forms appear, and then the cells may become rather indistinct. Increase in length of the filaments is by simple division, and new growths are produced by cells which break out of the sheath and

---

PLATE 1. *Sewage fungus.* **A** & **B** Two common growth forms of *Sphaerotilus natans* attached to pieces of grass. **C** *Fusarium aqueductum* growing on a dead twig. **D** *Apodya lactea* (*Leptomitus lacteus*) growing on a piece of moss. **E** *Beggiatoa alba* attached to lump of mud. **F** *Asellus aquaticus* overgrown with a coat of *Zoogloea ramigera* and *Sphaerotilus natans*. **G** *Nemoura erratica* (a stonefly) nymph overgrown with a coat of *Sphaerotilus natans* in which many vegetable fibres are entangled. **A–E** natural size. **F** & **G** x 5. All the specimens shown were preserved in 5% formaldehyde.

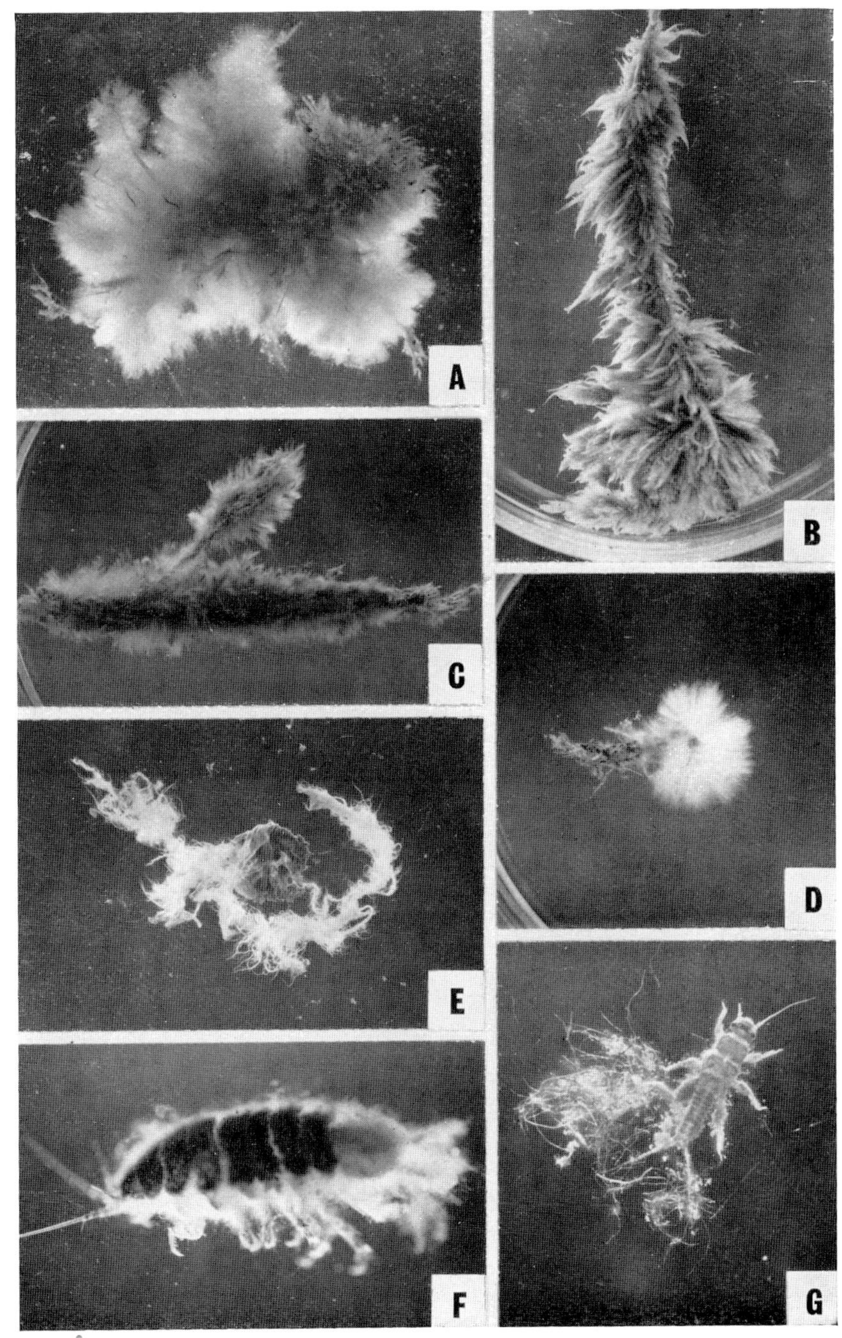
A
B
C
D
E
F
G

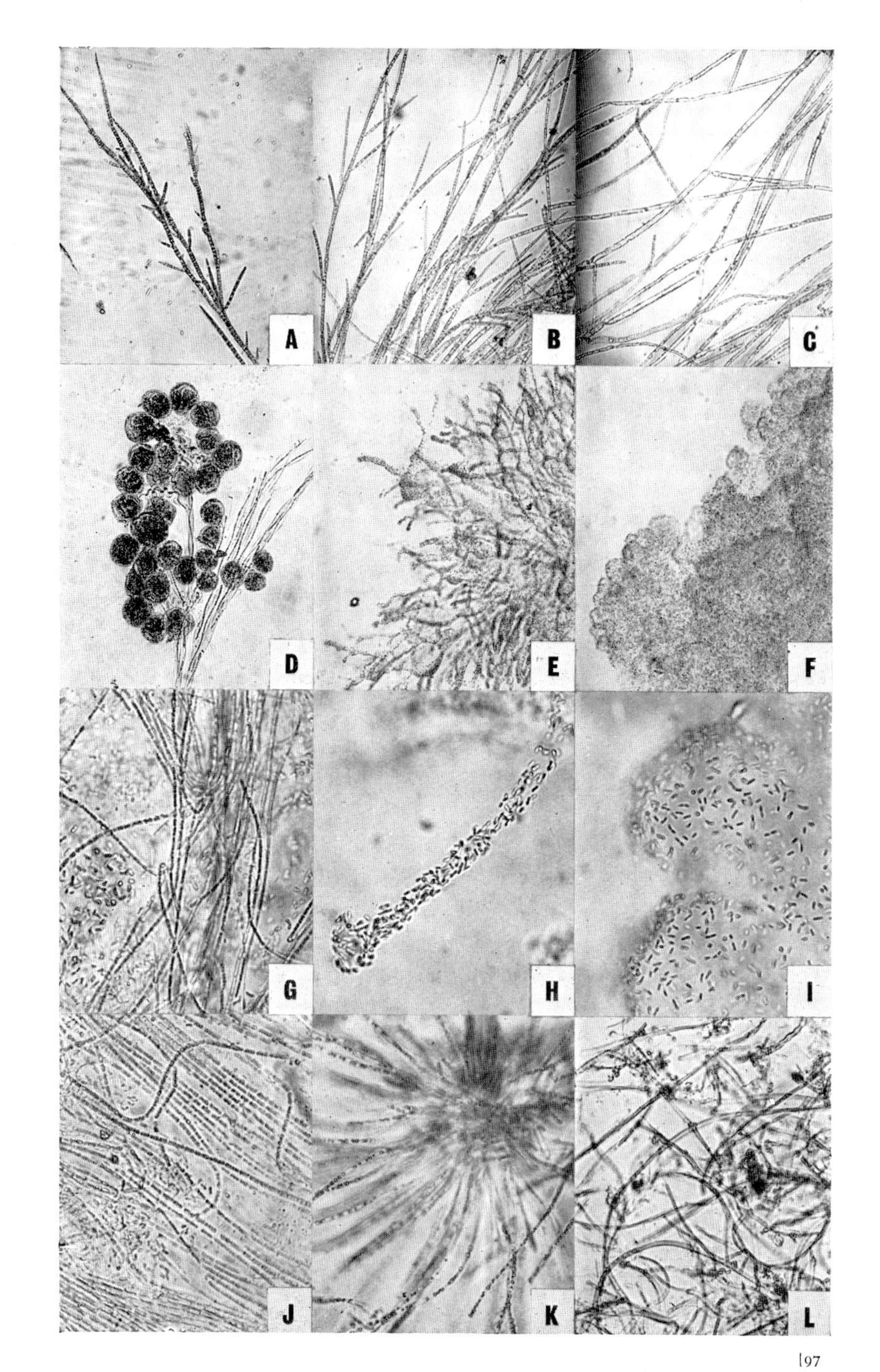
A
B
C
D
E
F
G
H
I
J
K
L

acquire flagella. They then swim away and settle down to form new filaments (Lackey and Wattie, 1940). This species grows satisfactorily only in running water, probably because it is there kept free of other bacteria which smother it in still water, but it is otherwise very tolerant and it grows in a wide range of pH and temperature. It is definitely aerobic, but it will grow in very low oxygen-concentrations. It cannot, however, withstand salt concentrations of more than about 300 mg./l., so it is unable to grow in brackish water. It feeds on organic matter, particularly carbohydrates, but it also needs nitrogen, which it can obtain from both organic and inorganic sources. It therefore thrives where there are amino-acids resulting from protein breakdown, especially where these are mixed with carbohydrates as occurs below sugar factories, breweries and dairies (Liebmann, 1953). It grows most rapidly at fairly high temperatures (25–30° C.), but it also grows well at temperatures as low as 6° C. (Butcher, 1932a). It thus tends to be particularly noticeable in winter when it not only extends further downstream, because of the slower decay of the organic matter, but is also largely freed from competition for foodstuffs with other bacteria, which are more affected by the cold. As it grows it entraps silt in large quantities (Butcher *et al.*, 1931) and its colour changes from whitish to brown. In general therefore the depth of the colour increases as one passes downstream to regions of lower organic content. The zone of active growth is, however, often very long, and where the basic foodstuff is fairly stable, as for instance are the carbohydrates in sugar-factory effluents, the *Sphaerotilus* carpet may extend for many miles below the outfall. Sewage and milk wastes are more readily attacked by other bacteria and they therefore produce shorter *Sphaerotilus* zones.

Where conditions for growth are particularly good, as, for instance, in sewage filters or near to an effluent outfall, *Sphaerotilus* produces another growth-form in which the cells are distributed at random in a branched mass of jelly, from which ordinary filaments sometimes project. This is known as *Zoogloea ramigera*, and it often occurs together

---

PLATE 2. Microphotographs of sewage fungus and other micro-organisms. **A** *Stigeoclonium tenue*—a green alga. **B** *Fusarium aqueductum*. **C** *Apodya lactea*. **D** *Carchesium polypinum*, as preserved in formalin—only one individual, in the middle on the right, is fully expanded. **E** *Zoogloea ramigera*. **F** The massive form of *Zoogloea* which occurs on sewage filter beds. **G** Filaments of *Sphaerotilus natans* mixed with *Zoogloea*. **H** Part of **E** under high power magnification. **I** Part of F under high power magnification. **J** *Sphaerotilus natans*. **K** *Beggiatoa alba*—note the granules of sulphur in the filaments. **L** *Leptothrix ochracea*—the spirally twisted filament is a specimen of *Gallionella ferruginea*, another iron bacterium. **A–F** x 60. **G–L** x 360. All the specimens shown were preserved in 5% formaldehyde.

with massive *Zoogloea* colonies of cumulus cloud-like form. The latter type of growth, which is illustrated in *Fig.* 17 and *Plate* 2, is the one usually found in sewage filters, where it is the most abundant micro-organism. This variation in growth-form indicates that the name *Zoogloea* probably embraces a complex of different types of bacteria. Butcher has shown (Butcher *et al.*, 1931) that in the river Lark, which was polluted by a sugar factory, there was a transition, as one passed downstream from *Zoogloea* to *Sphaerotilus* of various forms, to *Cladothrix*, revealing the change in conditions for growth.

Beggiatoa alba, one of the sulphur bacteria, is another constituent of sewage fungus. It forms a brittle white film on deposits of sludge (Bahr, 1953) and is common in the lower regions of bacterial filters (Wilson, 1949). The colonies consist of unbranched filaments of cells in which highly refractive deposits of sulphur can often be seen, particularly in older specimens (*Fig.* 17) (Butcher, 1932a; Liebmann, 1951). It occurs where hydrogen sulphide and oxygen are both present and oxidises the former to the element sulphur. It is therefore found only where reduction of sulphate to hydrogen sulphide occurs near to a source of oxygen, that is, for instance, on the mud surfaces which are the boundary between reduced and oxygenated conditions. Unlike *Sphaerotilus* it is tolerant of salt, and it is more common in brackish water than in fresh water, probably because sea water is very rich in sulphates.

In rivers in this country it seems to be unimportant, but its occurrence in large amounts would give a clear indication of the presence of sulphides, or of sulphur-containing proteins. Liebmann (1951) states that it occurs together with *Sphaerotilus* in the effluents from breweries, cellulose factories and dairies, and it is also common in waters polluted by wood-pulping factories (Vallin, 1958). This is undoubtedly because both the commonly used methods of digesting wood employ sulphur-containing salts (Pehrson, 1958), some of which escape into the effluent. *Beggiatoa* is only the most obvious of the sulphur bacteria, of which there is a whole range of species (Liebmann, 1951; Postgate, 1954), many of which are coloured and form red or green slimy coats on mud surfaces. Most of these species are anaerobic and occur only where there are high concentrations of hydrogen sulphide. They are therefore found more frequently in stagnant brackish pools than in rivers, but they have been reported from very foul streams (Berg, 1943). There are also a number of species of *Thiobacillus* which carry the process of oxidation a stage further by oxidising sulphur, or the thiosulphates in gas liquor, to sulphate. In so doing they produce sulphuric acid which may damage concrete or iron structures. Because of this, one species has received the name *Th. concretivorus*—the concrete-eating sulphur-bacterium (Postgate, 1954).

Apodya lactea, often also known as *Leptomitis lacteus*, which is a true fungus and a member of the Phycomycetes, is occasionally an important constituent of sewage fungus, but it seems to be rather rare in Britain. Its growths resemble those of *Sphaerotilus*, but they are not slimy and are more like cotton-wool. Under the microscope the filaments, or hyphae, can be seen to be much wider than those of *Sphaerotilus* and they are constricted at intervals (*Fig.* 17). They are also, unlike the hyphae of *Fusarium* (see below), quite undivided by cell walls. *Apodya* occurs where there is an ample supply of oxygen, calcium and nitrogenous organic matter of high moleculer weight (Stjerna-Pooth, 1957). It is therefore commonest where there is a very great dilution with hard

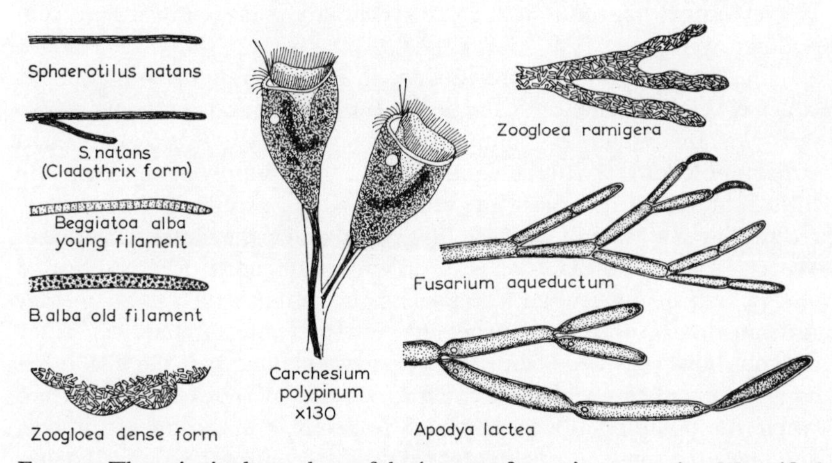

FIG. 17. The principal members of the 'sewage-fungus' community. Magnified x 330 except *Carchesium*

water. This may be why it is more important on the Continent than it is in Britain, where most of the rivers are comparatively small (Butcher, 1932a). Below wood-pulping factories it often largely replaces *Sphaerotilus* (Vallin, 1958; Mossevitch and Gussew, 1958), and, like *Sphaerotilus*, it is more obvious in winter (Liebmann, 1951). This is possibly because when the water is cold the rate of decay of the organic matter is slower and the oxygen content higher. Conditions are therefore more favourable for *Apodya*, and it is able to compete successfully with *Sphaerotilus*.

Fusarium (*Nectria*) *aqueductum* is also a true fungus, an Ascomycete, and like *Apodya* it is much less common than *Sphaerotilus*. Its hyphae are divided into cells by transverse walls, and they sometimes end in crescent-shaped spores (*Fig.* 17). Macroscopically the colonies resemble the other species, but they are brick-red rather than white or greyish. This species seems to favour rather acid waters (Butcher, 1932a; Vallin,

1958), and to grow only where there is a good supply of oxygen: its occurrence in sewage-polluted water is therefore rare, and it could be loosely described as an 'industrial sewage-fungus'.

A few other species of true fungi also occur in organically polluted water, including species of *Mucor* (Liebmann, 1951; Vallin, 1958), *Geotrichum*, *Penicillium* (Cooke, 1954) and *Achlya* (Stjerna-Pooth, 1957), but they do not often form massive growths as do the two species described above. Doubtless further study would reveal that, as with animals and green plants, there is a considerable amount of correlation between the occurrence of different types of fungi and the general condition of the water. Little work has, however, been done on this problem although Harvey (1952) has found that some species are persistently absent from polluted water.

The last important constituents of sewage fungus are animals, colonial bell-animalcules of the genus *Carchesium* and occasionally also *Epistylis*. In organically polluted water, particularly where there is a large population of bacteria together with a good supply of oxygen, these little Protozoa form large white colonies, which may be 2–3 mm. long, resembling the tufts of *Sphaerotilus*, with which they are often mixed. We have seen that these animals are an important constituent of activated sludge, and below sewage works using this process their colonies may form an almost continuous carpet on every solid object in the river. They cannot, however, withstand total de-oxygenation, so, like the other constituents of this biological community, they do not occur in reaches where the pollution is severe enough to result in anaerobic conditions, and they are usually most abundant at the lower end of the sewage-fungus zone.

None of these organisms is confined to polluted water; they all occur elsewhere under quite natural conditions, but the presence of organic pollution affords them opportunities for massive development. Given the right conditions for growth they appear with great rapidity, and though a river bed may be swept apparently clean by a flood the ragged masses reappear in a few days. Butcher showed that clean slides immersed in the river Lark, below the outfall of a beet-sugar factory, were colonised by *Sphaerotilus* cells to the extent of 25/sq. mm. after only 24 hours, and that after 4 days the filaments were over 1 cm. long and were beginning to collect silt particles. Thereafter the volume of the growths increased steadily for about three weeks, and the amount of the trapped inorganic matter increased until it was well over 90 per cent. of the total dry weight (Butcher *et al.*, 1931). These growths therefore encourage the deposition of silt where it would not otherwise collect, and they aggravate the effects of suspended solids in the effluent. They form a continuous blanket over eroding substrata, and may do the same on mud surfaces. They also seem to be repellent to fishes and other animals, which disappear from the areas where they grow even when it can be shown that the water is neither toxic nor severely de-oxygenated (Stundl, 1958). Some of this effect is undoubtedly caused by the change in substratum, but it has been shown that water fleas (*Daphnia magna*) are smothered by loose filaments in the water (Mossewitch and Gussew, 1958), and possibly fishes are repelled by the filaments adhering to their gills.

As the colonies age and break up, or are torn away by floods, they form a light flocculent suspended matter which drifts away and settles further downstream. In this way, even if the only primary effect of an effluent is to produce sewage fungus, it causes a blanketing of the river bed for a long distance, with corresponding effects on the plants and animals.

Algae

The algae of organically polluted waters have been extensively studied by Butcher, who has summarised his findings in two papers (1946b, 1947). His work was, however, all based on the algae which grew on microscope slides. Other workers (e.g. Kolkwitz, 1950; Liebmann, 1961; Fjerdingstad, 1950; Mack, 1953; Blum, 1957) have worked on scrapings taken from the river bed, and have reported on rather more species than has Butcher, although their work was less detailed and not numerical. We have already seen that the slide technique has both advantages and disadvantages over other methods of investigation. I shall attempt here to combine the results obtained by the two techniques into a general account, but it should be stressed that for all numerical data we are indebted to Dr. Butcher.

Where organic pollution is so severe as to cause total de-oxygenation it results in the elimination of all algae. If, however, some oxygen remains and a zone of sewage fungus is produced below the outfall, the algae are at first reduced in numbers and then increased further downstream. Examples of this effect are shown in *Fig.* 18, which is based on tables given by Butcher (1947). The numbers of algae growing on slides are plotted against distance downstream. The graphs have been so arranged that the source of pollution is just below mile 0, but the river Tame was receiving several effluents during the first 3 or 4 miles. It can be seen that, in the Trent, the numbers fell at first and then rose to a peak 8 miles below the outfall, and this peak was followed by another 25–30 miles downstream of the outfall. It is unknown if this second peak, well below the sewage-fungus zone, is an unusual occurrence; it was not observed in the Tame, where there was only a single peak. In the Bristol

Avon, where the pollution was very mild, the initial decrease did not occur, and the numbers of algae rose very rapidly to a peak, which also declined rapidily.

It is reasonable to suppose that the increase in numbers of algae at

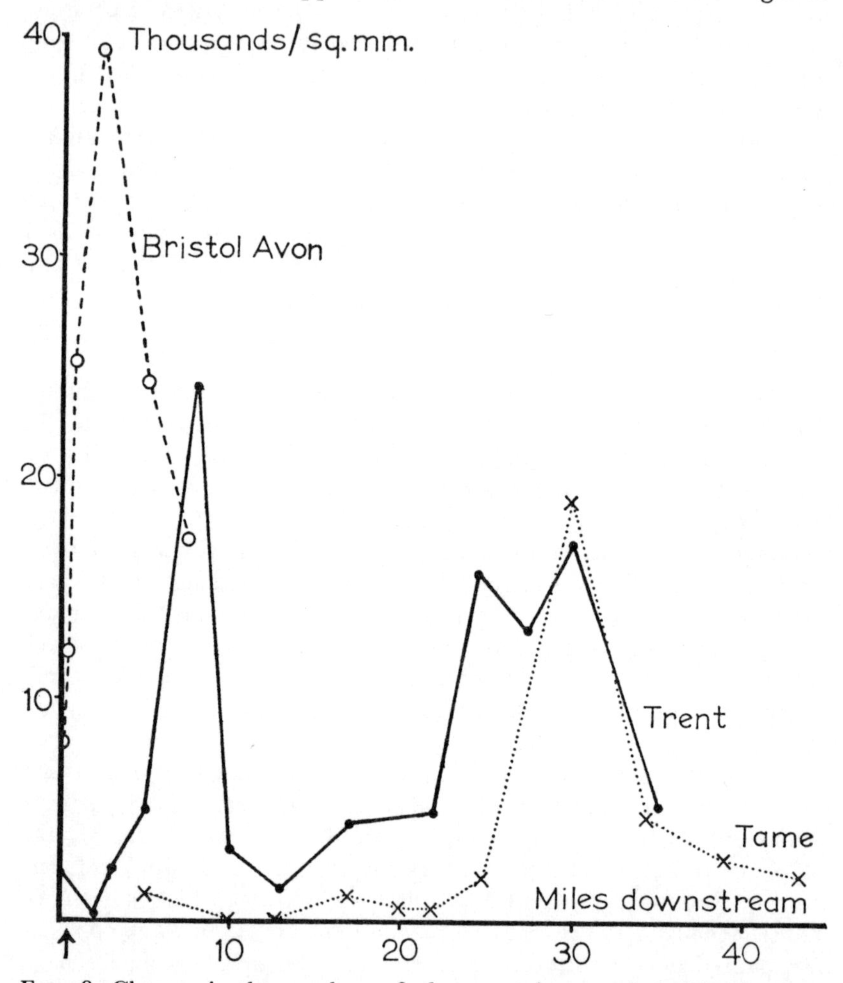

FIG. 18. Changes in the numbers of algae growing on glass slides in three organically polluted rivers. Data plotted from Butcher (1947). The arrow shows approximately the position of the outfall.

points downstream of the outfalls was due to the release of nutrient salts from the organic matter, and it is probable that the different positions of the peaks are the result of the different types of organic matter involved. The Avon was polluted by milk wastes, which are very readily decomposed, while the pollution in both the Trent and the Tame was caused by a complex mixture of effluents. Very probably the double peak which occurred in the Trent was the result of this complexity. The numbers of algae do not remain high for a long distance downstream for a variety of reasons; the nutrients are used up by the proliferation of algae, they become more diluted as the size of the river increases, and large populations of algae are unstable. Also, as one proceeds downstream into zones of cleaner water the number of normal grazing river-invertebrates increases and these keep the algae under control.

In addition to the fluctuations in numbers of algae described above, the species also change (*Fig.* 4). The normal eutrophic algal flora of a wide variety of species dominated by *Cocconeis*, *Ulvella* and *Chamaesiphon* disappears: several authors have noted, for instance, that the number of diatom species is reduced by pollution (Patrick, 1954; Westlake and Edwards, 1956). But some algae are particularly resistant; amongst sewage fungus one can often find filaments of the blue-green algae *Oscillatoria* and *Phormidium* (Liebmann, 1951; Fjerdingstad, 1950; Mack, 1953), the green alga *Ulothrix* and the flagellate *Euglena viridis*; and particularly near the water surface, as on the tops of large stones, dense bright-green colonies of *Stigeoclonium tenue* (*Fig.* 4, *Plate* 2), which contrast sharply with the grey and white of the sewage fungus, occur. Similar phenomena are reported from America (Campbell, 1939; Brinley, 1942; Blum, 1957). These are all algae which are characteristic of water which is very rich in nutrient salts, and it appears that *Stigeoclonium tenue* thrives particularly where there are good supplies of nitrate (Westlake and Edwards, 1956). The above are algae which are readily visible to the naked eye when present in large amounts. Butcher's (1946b) detailed studies showed that two small diatoms, *Gomphonema parvulum* and *Nitzschia palea*, are also very resistant to organic pollution, and occur in large numbers on slides even in the sewage-fungus zone, and they may be accompanied by *Navicula* species and *Surirella ovata* (Mack, 1953; Blum, 1957).

Fig. 19 which is based on tables given by Butcher (1947), shows details of the observations made on the rivers Tame, Trent and Bristol Avon; similar data are reported from the river Churnet above the point where it was polluted by copper (Chapter VII). It will be seen that the normal *Cocconeis*/*Chamaesiphon*/*Ulvella* community disappeared for a long distance below the source of pollution, except in the Avon, where pollution was slight, and that it was replaced, or in the Avon supplemented, by *Gomphonema*, *Nitzschia* and *Stigeoclonium tenue*. These three algae were the main constituents of the peaks in the Tame and Trent, but in the more mildly polluted Avon the normal species also increased.

Further downstream a succession occurred: the normal community reappeared and at the same time two other species of *Stigeoclonium* appeared and persisted for a while. The species in this genus are not well defined (Butcher, 1947) and it may well be that all three are growth-forms of a single species, *S. tenue* being merely the one which appears in very enriched water.

Finally, as recovery is completed, the normal flora is re-established, the numbers fall to more usual levels, although they may remain higher than they were above the outfall, and the three types which were par-

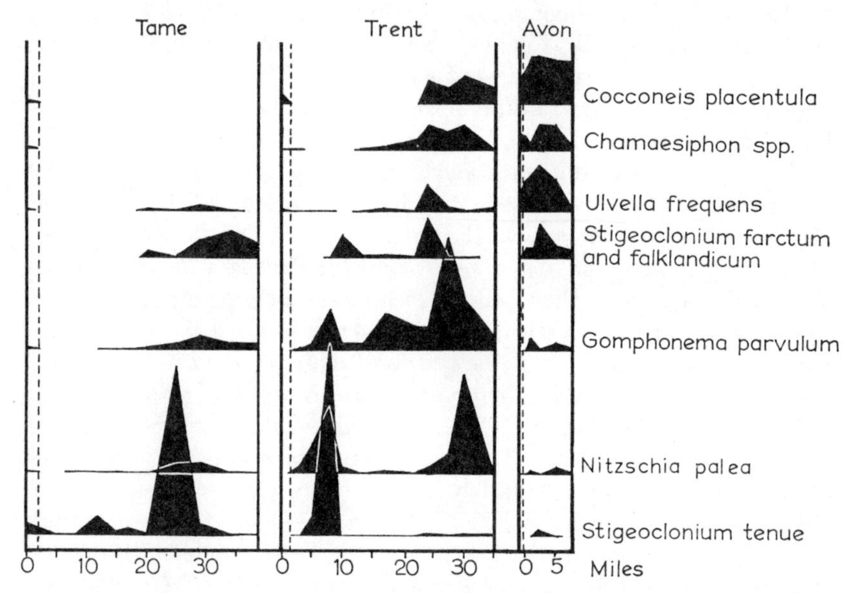

FIG. 19. Changes in the composition of the principal members of the algal flora growing on glass slides in three organically polluted rivers. Data plotted from Butcher (1947). The vertical dashed lines show approximately the positions of the outfalls.

ticularly abundant in the polluted zone return to their normal status of occasional occurrence. It should be stressed, however, that this return to normal requires a long distance: 43½ miles from its source the Tame had not recovered its normal flora, after 35 miles the Trent was only approaching its original condition, and even in the Avon the effect of the pollution was still clear 7½ miles downstream.

In this discussion of the effect on the number of species we have of course only been considering those species which are dominant on submerged slides. In unpolluted regions where *Cocconeis, Ulvella* and *Chamaesiphon* thrive there are also many other species. These also begin to reappear, and the list of species becomes longer as one proceeds further into the zone of recovery (Fjerdingstad, 1950; Liebmann, 1951). Indeed, although there are a few algae, such as *Draparnaldia* and *Lemanea*, which seem not to recur below severe organic pollution, the general enrichment of the river water by the breakdown of the organic matter probably results in an increase in the number of species present. Blum (1957), for instance, reports finding large colonies of *Spirogyra*, *Closterium*, *Euglena*, *Tetraspora* and *Diatoma* in quiet reaches of a polluted river in Michigan. As, however, the distance required for recovery is so long it would be difficult to disentangle this effect from the general change in the character of the river over a long distance. So far as I am aware no special study has been made on this aspect of the algae of rivers, but, as we shall see, it is an important point in the pollution of lakes.

So far we have been considering only the smaller algae, but one species, *Cladophora glomerata* or blanket-weed (*Fig.* 4), often develops massive growths in polluted rivers. These resemble large, wet, green blankets, and are present only in the summer, when they may completely cover the river bed. In the autumn, or at times of flood, they break off and drift downstream, where they may cause trouble by lodging against the piles of bridges and blocking the water course. If they enter lakes or still reaches on the course of the river their decay may result in deoxygenation or nuisance: decaying *Cladophora* looks, and smells, rather like untreated sewage. The dense and rapid growth of this alga smothers the normal river weeds (Liebmann, 1951), and in the river Lee in Hertfordshire I have observed drifting masses becoming entangled with plants which were then either defoliated or uprooted. Another undesirable property of large masses of *Cladophora* is that at night, when they are not photosynthesising and producing oxygen, they tend to de-oxygenate the water by their normal respiratory process (Butcher *et al.*, 1937). Other aquatic plants also, of course, have the same effect, but to a much lesser extent, as they do not occur in such massive amounts.

Cladophora glomerata is a common plant of eutrophic water, and it often occurs in ornamental ponds and aquaria as well as in rivers. In unpolluted rivers its growths are short, densely tufted and pale green, but in water which has been organically polluted they may be 2 feet or more long, the filaments are richly branched and their colour is dark green (Butcher *et al.*, 1937). Such growths appear below the sewage-fungus zone and may extend for many miles. Butcher (1933) suggests that they occur only where some organic matter is present and the amount of dissolved nitrate is high, but later work (Butcher *et al.*, 1937; Westlake and Edwards, 1956) suggests that a rich supply of nutrients is the most important factor. This would seem to be probable because, in my experience, massive development of blanket-weed occurs where the B.O.D. is falling rapidly (*Fig.* 16) and the completion of mineralisation is releasing nutrient salts. The growths do not diminish as the B.O.D. falls, but extend into reaches where its level is quite low but where the concentration of nutrient salts is still relatively high.

During the winter this alga apparently disappears, but small basal portions remain on fixed objects and give rise to new filaments in the spring. These then reproduce by flagellated spores, and the whole river bed is rapidly recolonised. The large growths remain until the autumn unless swept away by floods, and they entrap much silt and vegetable debris.

Butcher observed that in the river Tees the amount of *Cladophora* varied very much from year to year. This sort of variation may be due to climatic factors, but it may also be caused by variations in the quality and amount of the organic matter entering the river. In the river Lee I observed changes in the *Cladophora* zone which could be correlated with the general quality, as assessed by chemical analysis, of the effluent from a sewage works, which is the major source of the water in the river. In 1950 the *Cladophora* zone began 2 miles below the outfall and extended for at least 7 miles. In 1952, after a year in which for various reasons the average levels of B.O.D. and suspended solids of the effluent had been particularly low, this zone began actually at the outfall and extended for 12½ miles.

In 1953 and 1954, however, when the quality of the effluent had become stabilised at slightly less satisfactory levels, the *Cladophora* had again moved downstream, and it extended from 1¼ to 13¼ miles below the outfall. Here we have a clear indication that this plant does not grow satisfactorily until a certain state of mineralisation has been achieved. In 1952 this level was being reached in the sewage works, but in 1950 the effluent needed 2 river miles to reach the required level, and in 1953 and 1954 1¼ miles. In all years the distance over which the large growths extended was probably about the same, but their lower limit was not determined in 1950.

Higher Plants

Rooted plants and mosses are eliminated by gross organic pollution as they are smothered by silt or sewage fungus, but when the pollution is less severe and the water is not rendered too turbid for photosynthesis a few species of macrophyte may survive. The commonest of these is the pond-weed *Potamogeton pectinatus*, which may on occasion be found even in the sewage-fungus zone. We have seen that this plant is a normal member of the silted community, and doubtless the excessive silting of the stream bed favours its development. Unlike most of the other members of this community it has long smooth grass-like leaves, which are readily kept clear of suspended matter by the current and for some reason they are not a suitable substratum for *Sphaerotilus*.

In my experience the only other macrophyte which occurs together with sewage fungus is the moss *Fontinalis antipyretica*, but only in very shallow riffles where the force of the current keeps the stones free of silt, and the moss is always much overgrown with sewage fungus. Below the lower limit of active growth of sewage fungus but above the *Cladophora* zone I have found the moss *Eurhynchium riparioides*, but, like *F. antipyretica*, only in shallow riffles where the current keeps it free of silt and drifting fungus. In deeper, more sluggish water one may sometimes find considerable amounts of *Potamogeton natans*, a species which normally occurs in ponds. It seems possible that this plant is less affected by a constant rain of silt and increased turbidity of the water than are the other broad-leaved water weeds, because its leaves float on the surface like those of the water lily: it is, however, like the water lily, able to tolerate only very slow currents.

Further downstream, where suspended solids are reduced, the normal communities of river weeds reappear, although they may be damaged by *Cladophora*. Usually, however, the effect of organic matter is to make the water less permeable to light, because of discoloration or increased turbidity, with the result that plants are confined to shallower water. Set against this, however, is the increased fertility, which encourages plant growth: much therefore depends on the detailed topography of the river and the nature of the effluent.

In addition to the fairly readily explainable effects outlined above there are a few indications that organic effluents may be toxic to some plants, of which starwort, *Callitriche*, appears to be one. It was been shown that *Potamogeton densus*, a member of the partly silted community, is killed by 14 days' exposure to 6 p.p.m. of synthetic detergent (Ministry of Housing, 1956): other plants, however, including Canadian pond-weed *Elodea*, are unaffected by much higher concentrations (Degens *et al.*, 1950). Similarly some members of the emergent community, particularly the tall grass *Glyceria maxima*, become yellowed where they grow along the banks of rivers polluted by sewage, the plants further out into the water being more severely affected than those nearest to the bank. The reason for this is unknown, but it has been suggested that it may be caused by heavy metals, which are always present in the suspended solids of sewage (from pipes, boilers, ointments, toothpastes, etc.) and which accumulate round the roots of the plants. This aspect of organic pollution clearly needs further study: but so also do most of the others.

Protozoa

The micro-fauna has often been investigated with special reference to organic pollution, and a large number of species are described by Liebmann (1951) as characteristic of certain degrees of sewage pollution. Unfortunately these organisms are very difficult to identify, and Liebmann and other workers (Lackey, 1938; Mohr, 1952) have stressed that accurate identification to species is essential, as different species in the same genus react quite differently. As this can very often be done only on living specimens, examination has to be made very soon after the collection of samples. Other workers have stressed that the numerical abundance of the various species is as important as their occurrence, and that no biological investigation which does not take this into account is reliable (Šrámek-Hušek, 1958).

Study of the very small animals therefore presents formidable technical difficulties, and it is not surprising that the various published results are not in very close agreement. There is also a serious difficulty in the method of sampling. Most of the Protozoa which give particular indications of pollution live in or very near the river bed (Mohr, 1952) or amongst the sewage fungus. They are therefore difficult to sample quantitatively although methods for doing this in a comparative manner have been developed (Šrámek-Hušek, 1958). Several investigators have therefore confined their attentions to the planktonic inhabitants of the open water, which are much easier to sample, but the results of their studies have been very conflicting. In the Mississippi (Wiebe, 1927) and the Vistula (Turoboyski, 1953), both of which showed clear signs of organic pollution, as shown in their bottom faunas and water chemistry respectively, samples above and below the sources of pollution were found not to differ.

On the other hand Lackey (1938, 1941, and in Phelps, 1944), after an extensive study of the plankton of the Scioto river, a tributary of the Ohio, was able to show that some species became more abundant below a source of sewage pollution, while others declined in abundance. But the changes were not very marked and his results are not conclusive despite a very thorough investigation. Brinley (1942), summing up the study of plankton in the Ohio basin, remarks that 'the plankton population gives a better picture of the past history of a certain volume of water than it does of the sanitary conditions of the stream at any definite location'. This is, indeed, what one would expect, because any particular water-mass will travel fast or slowly according to the river level at the time, and the slow cycle of changes from protozoans which are primarily bacterial feeders to those which have other feeding habits will proceed without reference to the actual position of the water. This point has also been stressed by Šrámek-Hušek (1958).

Another difficulty is introduced by the fact that the soil is full of Protozoa, and these are continually washed into rivers. Gray (1955) made a detailed study of a polluted English stream and showed that while the largest numbers of ciliate protozoans, and of species, were associated with algae, it was probable that they feed there on the bacteria introduced into the water by the pollution. The numbers, however, fluctuated with the weather, being high after continuous drought, which allowed them to multiply without being swept away. They also increased after heavy rain, and such increases were too sudden to be accounted for by reproduction, and must have been due to colonisation from the soil.

With all these complications and contradictory results it will be appreciated that it is impossible at this stage in our knowledge to give a satisfactory account of the effects of organic pollution on microscopic animals, and that most of the published investigations can refer only to the particular conditions which were being investigated. It is, however, clear that, when organic pollution is very severe, only bacteria and a few flagellates such as *Bodo* occur. As conditions improve a little, and allow the development of *Sphaerotilus*, bacteria-eating Protozoa, amongst which *Colpidium colpoda*, *Glaucoma* and *Paramecium* species are important, appear in large numbers. With further improvement these are joined by *Carchesium* and *Vorticella* and such algal-eating genera as *Chilodonella*, *Spirostomum* and *Stentor*, and then by an increasing number of types including *Coleps*, *Didinium*, *Lionotus* and many others, together with Rotifers.

At first then the microscopic animals are few in kind but abundant in number and all are bacteria feeders, and their variety and diversity of feeding habit increase and their individual abundance declines as conditions improve. Most of these creatures, some of which are illustrated in *Fig.* 20, are to be found in sewage filters (Wilson, 1949) and in activated sludge (Sládeček, 1958), as well as in polluted rivers. Finally, as mineralisation nears completion, large numbers of green flagellates such as *Euglena* and *Phacus* may appear. For further details the reader is referred to Lackey (1938), Thomas (1944), Liebmann (1951) and Šrámek-Hušek (1958), but with the warning that in some respects the lists of these authors do not agree.

Obviously our knowledge of the real significance of many of these creatures in the study of pollution is still very far from complete, and numbers are probably at least as important as the types which occur. As with all organisms which abound in polluted water, all the species may occur in places where there is no pollution. Once again we meet an aspect of this subject which is clearly in need of much further study. At present there is no really satisfactory technique which will collect suitable numerical samples of the microscopic animals inhabiting river beds, especially when one takes into account the great variation of microhabitat which such environments offer. Possibly some method similar to the slide technique used by Butcher for the algae would give satisfactory results, but all examination of samples would have to be made on living material. It seems fairly clear that samples of plankton alone are less

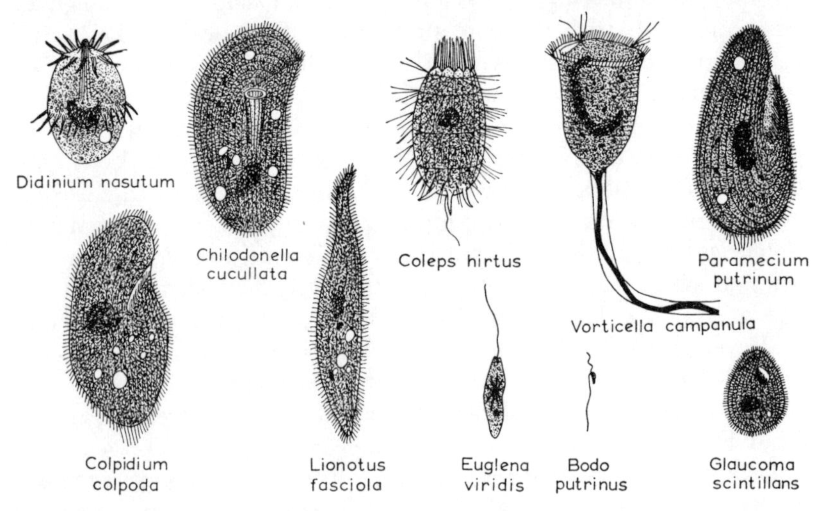

FIG. 20. Protozoa which are important in the study of organically polluted water. Magnified x 150.

likely to yield much further knowledge on the problems of the biology of pollution, although they may prove to be useful in tracing the history of a particular water mass. I should imagine, though, that chemical analysis would give equally satisfactory results with far less work and without the necessity of the services of a highly specialised scientist.

Larger Invertebrates

In considerable contrast to the microscopic animals the larger invertebrates show very clear reactions to organic pollution, and most of them are much easier to identify. The faunal changes induced by such pollution have been found to be very similar on both sides of the Atlantic, despite the differences in species of larger aquatic invertebrates which inhabit Europe and North America. There is much less information from other parts of the world, but similar findings are reported from South Africa (Harrison, 1958b), and it seems probable that the same general type of succession occurs everywhere. On the face of it this is surprising because the species of larger animals, unlike the Protozoa, algae and bacteria, have strictly limited geographical distributions: Europe and North America for instance share rather few species of larger aquatic invertebrates, and very few of these occur also in South Africa. However, when pollution is fairly heavy the effects on most invertebrates are so marked that whole taxonomic groups, rather than individual species, are involved, and it is only when pollution is very mild that actual specific differences become important. This therefore makes the larger invertebrates more useful as indicators of pollution than are the Protozoa, quite apart from the fact that they are readily identifiable when preserved. Amongst the ciliates different species of the single genus *Vorticella* react very differently to gross pollution, the same applies to species of *Amoeba*, *Paramecium* and *Trachelomonas* (Lackey, 1938; Liebmann, 1951). Amongst higher invertebrates differences in reactions of the several species within a single genus seem to be less marked, and thus very detailed identification is not essential in order to obtain a general impression of the conditions, although it becomes necessary, as we shall see, in the study of mild pollution.

Where the concentration of organic matter is high enough to produce total de-oxygenation no normal river animals survive, but they may be replaced by the larvae of the moth-fly *Psychoda* and the bee-fly *Eristalis* (*Fig.* 21). These larvae are both air-breathers whose normal habitats are rich in organic matter and so often de-oxygenated, and they overcome the problems which such conditions present by breathing through air tubes which open at the tips of their tails. Thus they need only to keep their tails at the surface while the rest of their bodies are immersed in de-oxygenated water. *Psychoda* normally inhabits mud flats (Lloyd, 1944) and the edges of ponds, and has found very suitable conditions for life in man-made habitats such as drains and sewage filters, from which the flies often emerge in clouds. The larva of *Eristalis* is the rat-tailed maggot, which has an extensible telescopic tail enabling it to descend 3 or 4 in. into the water without losing contact with the atmosphere. Its normal habitats are foul ponds and other places where there is much decay, such as pools fouled by cattle dung and the tubs used for preparation of liquid manures: it often occurs in large numbers along lake shores where the wind has piled up large masses of rotting water weeds.

Neither of these flies, nor the mosquito *Culex* (*Fig.* 21) which sometimes occurs with them (Liepolt, 1953) and also has an air-breathing larva, is a normal river-animal and, although they have often been reported from grossly polluted rivers (Kaiser, 1951; Gaufin and Tarzwell, 1956), they can live only along the banks where the water is shallow and they are not carried away by the current and drowned. They are occasionally to be found in such situations even where the pollution is not so severe as to cause septic conditions, and it may be that such populations of *Psychoda* are maintained by constant immigration from the biological filter-beds (Kaiser, 1951). Mosquito larvae are presumably favoured by the rich supply of particulate organic matter on which they feed.

Where only very little oxygen remains in the water, or the river bed

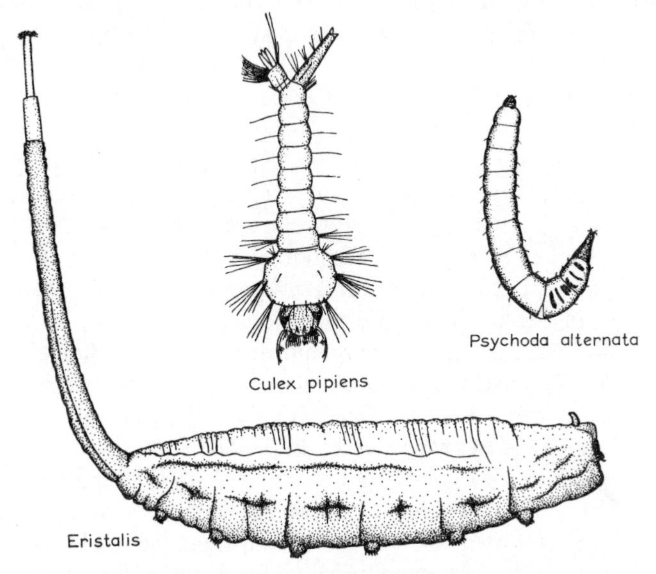

FIG. 21. Fly larvae which may occur in the septic zone of grossly organically polluted rivers. Magnified x 5.

is completely covered over with organic solids or sewage fungus, the main inhabitants are always the so-called sludge worms of the family Tubificidae. Unfortunately little detailed work has so far been done on these worms, and we do not yet know just how many species are regular inhabitants of polluted water nor whether the different species are dominant under different conditions. Species of the genera *Tubifex* (*Fig.* 10) and *Limnodrilus* have been most frequently reported from grossly polluted water, but others may occur, and these two genera are of course by no means confined to foul water. In such places, however, they are particularly favoured; they have abundant food in the rich organic mud, and they are free from enemies and competitors which cannot stand the low concentrations of oxygen. They therefore often occur in enormous numbers; up to 350,000/sq. yd. have been recorded (Richardson, 1929) and their waving tails may make the river bottom quite red.

If the river is shallow and turbulent the water may be reasonably well-oxygenated even in the zone where there is a continuous and actively growing carpet of sewage fungus. Under these conditions Tubificidae may still be dominant in the accumulated organic matter, although accompanied by Chironomidae of various sorts (see below), but one usually finds a few clean-water animals also. For instance worms of the family Naididae (e.g. *Nais*, *Fig.* 11) are often to be found in dense growths of sewage fungus in shallow turbulent water, and also the shrimp *Gammarus pulex* (*Fig.* 6). Naids seem to thrive under such conditions and to reproduce actively, but *Gammarus*, although it survives and apparently grows quite well, is never so numerous as it is in clean water, and it does not breed satisfactorily. This is at once apparent when comparison is made between samples collected above the outfall and in the sewage-fungus zone: the proportions of young specimens and of egg-bearing females are always very low in the latter.

Although in such localities the amount of dissolved oxygen may be high because of good aeration, the normal river animals, which are of course those of eroding substrata, are usually absent or very sparse. Undoubtedly the coat of sewage fungus and settled solids makes conditions unsuitable for them, and it certainly interferes with their holdfast mechanisms, all of which depend on the presence of smooth clean surfaces. Also if one does find odd stonefly or mayfly nymphs they are often competely overgrown with sewage fungus. A stonefly nymph affected in this way is shown in *Plate* 1, and it will be obvious that any animal so encumbered is unlikely to be able to lead a normal life. I have observed this type of growth only on *Asellus*, on some sorts of insects and on snail shells; the Chironomidae seem to be immune and it is probable that, as with the plants, there are variations in the detailed structure of body surfaces, only some types of which offer suitable substrata for sewage fungus.

Returning now to the more usual condition where pollution results in marked de-oxygenation of the water, sewage fungus grows less actively as one passes downstream to reaches where the concentration of organic matter is lower. The Tubificidae then begin to be accompanied by chironomid larvae, and these become increasingly important as conditions improve. Most obvious among them are the large red 'bloodworms' which are the larvae of the midge *Chironomus*. In British literature the species which occurs most commonly in polluted water is usually referred to as *C. riparius*, but this is the same species as the one called *C. thummi* on the Continent, and probably also the same one as occurs in polluted rivers in North America. We need not enter here into the controversy as to which is the proper name: suffice it to say that Professor Thienemann (1954), who is the world's leading authority on these insects, calls the species *C. thummi*, and that we shall use that name here.

This midge larva (*Fig.* 10) is a common inhabitant of muds which are rich in organic matter, and unlike some other members of the genus it can tolerate fairly high concentrations of salt, sulphuretted hydrogen and ammonia. It therefore thrives in such places as foul seaside rock-pools and farmyard ponds probably because of freedom from enemies and competitors. In polluted rivers it is occasionally accompanied by the slightly less tolerant species *C. dorsalis* (Thienemann, 1954). Like the Tubificidae these larvae find conditions very suitable as they have a rich supply of food in the form of bacteria and organic solids, and they may build up to enormous numbers. As a result it is quite a common sight to see clouds of the midges dancing near polluted rivers on still, warm summer evenings. The clouds, which consist of male flies, gather over prominent objects such as hedges and hayricks and resemble columns of black smoke, for which they have often been mistaken. As each midge is only about a third of an inch long and the column may be 20 or 30 feet across and 100 feet high, it requires little imagination to appreciate the enormous numbers involved, especially when several columns are visible simultaneously. Berg (1948) reports that *C. thummi* is generally absent from places where the current is strong, but this has not been my experience; I have found it in very rapid streams where sewage fungus and accumulated solids have filled the interstices between the stones and so formed a suitable microhabitat.

We have seen (Chapter IV) that Tubificidae raise mud from considerable depths to the surface. They must therefore make a marked difference to the rate at which the deposited organic solids are oxidised, as they constantly turn them over. *Chironomus* is an even more efficient oxidiser of mud, as its tubes are filled with water which the larvae keep in circulation by undulating their bodies. It has been shown that they cause a downward extension of the superficial layer of oxygenated mud (Westlake and Edwards, 1957), and that the addition of larvae to organic mud increases its rate of oxidation. Doubtless, therefore, quite apart from any which they eat, the larvae of *Chironomus* cause the destruction of a great deal of organic matter, but in doing so they of course increase the local oxygen demand. Edwards (1957) suggests that they may also cause the spread further downstream of masses of settled solids. He observed that below a sewage works outfall the surface of the mud, which was composed very largely of *Chironomus* tubes, became lifted by gas bubbles and floated away. This occurred only in spring, when the rising temperature induced a mass emergence of the flies, and he suggests that at other times, when the mud is well populated with larvae, their movements prevent the lodging of bubbles produced by anaerobic decay in the deeper layers. Clearly this sort of effect of the seasonal rhythm of this species may have a considerable influence on the extent to which pollution affects a river.

What we may call the *Chironomus* zone may begin in the middle, at the beginning of, or even above the sewage-fungus zone, but it often extends far beyond the sewage fungus and it attains its maximum development where the fungus is sparse. Probably the controlling factor here is the oxygen content of the water. Tubificidae and sewage fungus tolerate very low concentrations of oxygen; *C. thummi* is less tolerant. One can readily demonstrate this by placing samples of both types of animal together in a closed bottle and allowing the oxygen content to fall: the *Chironomus* larvae are always the first to die. The exact relationships of these organisms in a polluted river depends therefore on local conditions.

C. thummi and *C. dorsalis* are not the only chironomids of the *Chironomus* zone. They are often accompanied by species of the genera *Psectrotanypus* and *Procladius* (Thienemann, 1954), carnivorous members of the sub-family Tanypodinae which feed on tubificids. One also often finds large numbers of small green or white larvae of the sub-family Orthocladiinae, especially where the water is fairly well oxygenated. The principal species of these are *Prodiamesa olivacea*, which lives in mud, and *Cricotopus sylvestris* (*Fig.* 12) (Thienemann, 1954): the latter is usually associated with plants, but it may also occur in bare mud. Even in the peculiar acid conditions which favour the growth of *Fusarium*, chironomids of the same sub-family may occur amongst the fungus (Vallin, 1958). These Orthocladiinae, like *C. thummi*, may also occur in enormous numbers, but the Tanypodinae are, as always applies to carnivores, much less abundant.

As conditions improve further downstream and one passes into the regions of algal maxima, higher oxygen content and reduced suspended solids, more and more species appear. Other types of chironomid come in, the importance of *Chironomus* declines, and a notable addition to the fauna is the water slater *Asellus aquaticus* (*Fig.* 10). This species, like the tubificids and *Chironomus*, may occur in enormous numbers even on eroding substrata, on which it is otherwise very rare, and in *Cladophora* beds it may be astonishingly abundant. Here again we doubtless have an example of an animal which is able to take advantage of local conditions and to build up large populations in the presence of abundant food.

Most European workers have recorded this '*Asellus* zone' (Berg, 1948; Butcher *et al.*, 1931, 1937; Pentelow *et al.*, 1938; Butcher, 1955, 1946b) and I have observed it in several British rivers. Most American workers on the other hand make no mention of it even where their studies have been concerned with reaches of rivers where, in Europe, one would expect to find *Asellus* to be important (Richardson, 1921, 1929; Wiebe, 1927; Gaufin and Tarzwell, 1952, 1956; Campbell, 1939).

Indeed the only American reference to a connection between *Asellus* and organic pollution which I have found is that of Bartsch (1948), where the species is given as *A. communis*. As, however, this paper deals in generalities and not with a particular investigation, it is possible that Bartsch was influenced by European findings and that there is actually no North American *Asellus* species which thrives under the conditions which favour *A. aquaticus*. It may be noted that the other common European species, *A. meridianus*, is apparently never associated with polluted water.

In the *Asellus* zone, as in that dominated by *C. thummi*, *A. aquaticus* is not usually the only important species. Molluscs are often very abundant, as are leeches, and an important predator is the alder fly, *Sialis lutaria* (*Fig.* 10), which is often quite common. As Pentelow has remarked, organic pollution often results in the replacement of the normal river fauna with one consisting of *Asellus*, *Sialis*, leeches and molluscs (Pentelow *et al.*, 1938). In America this community seems to consist primarily of leeches and molluscs, among which the mud-dwelling genus *Musculium* (*sphaerium*) is very important (Richardson, 1929; Wiebe, 1927).

In Britain the molluscs which may occur in abnormally large numbers in this zone are *Sphaerium lacustre* and *S. corneum* (*Fig.* 11) (which appear to succeed one another in importance as one passes downstream), *Limnaea pereger* (*Fig.* 7) and *Physa fontinalis* (*Fig.* 11). Important species of leech are *Trocheta subviridis* and *Erpobdella testacea*, (which seem to favour polluted water), *Glossiphonia complanata* (*Fig.* 5) and occasionally *Helobdella stagnalis* (Mann, 1952, 1953). I have also sometimes observed enormous populations of the small midge *Tanytarsus* (*Fig.* 10), whose larvae make tubes in deposits of the coarse silt which results from the mineralisation of suspended sewage solids. In the *Asellus*/*Cladophora* zone of the river Lee, silty parts of the river bed appeared to be made of the tubes of these insects, and I have observed this phenomenon in other rivers also. The larvae are small (*Fig.* 7) and inactive, and as they do not readily leave their tough tubes they are easily overlooked.

The distances over which the tubificid, *Chironomus* and *Asellus* zones, or the roughly corresponding sewage-fungus and *Cladophora* (or algal maximum) zones, extend vary greatly; and if the initial load of polluting matter is small or already well mineralised the upper part of the succession may be absent altogether. For instance in the river Lee, which was quoted as an example when we were discussing *Cladophora*, the pollution produced a short *Chironomus* zone in 1950, 1953 and 1954, but in 1952 this had disappeared and the *Asellus/Cladophora* zone began at the effluent outfall. At no time during these years was the tubificid/sewage-fungus zone present. It would seem that, unless dilution is very great, the factor which determines which zone shall appear first below the outfall is the actual nature of the organic matter. In the river Lee, although virtually the whole of the water in the river was sewage-works effluent, the treatment had been so thorough that the organic matter which remained could no longer serve as food for sewage fungus, and even when the *Chironomus* zone was present it was only a mile or two long. The *Asellus/Cladophora* zone was, however, 12–13 miles long, so the effects of the pollution produced obvious biological changes for about 15 miles.

On the other hand pollution by mill wastes in the Bristol Avon produced all the zones, but all within about 5 miles (Pentelow *et al.*, 1938) Here organic matter was discharged intact and served as food for sewage fungus, but it was so diluted and so rapidly oxidised that the biological sequence was soon over and, although the various members of the pollution fauna were encouraged, conditions were nowhere so bad that the fauna was reduced to Tubificidae. Indeed even in the short sewage-fungus zone, which was only about ¾ mile long in the summer, *Chironomus*, *Asellus*, leeches and a few other animals occurred with the tubificids. In milder cases, therefore, the zones are less distinct from one another as they overlap so much, and a few of the more tolerant river animals may persist right through them.

An even more striking example of very rapid passage through this biological sequence was observed on the Tees (Butcher *et al.*, 1937) where it is polluted by its small tributary the Skerne. In this river, probably because of great dilution and rapid re-aeration, recovery was complete and the invertebrate fauna back to normal, except for the persistence of leeches, after only 600 yards. However, not only the amount and the state of oxidation of the organic matter is important; its nature also influences the length of the zones. If it is readily oxidised, as were the mill wastes in the Avon, it soon disappears, but if it is very stable it will travel a long way before being completely mineralised. Butcher (1955) points out that in the river Trent the biological effects of pollution by sewage often extend for only 5 miles or so, but other more stable effluents, such as gas liquors, extend their influence much further. And, of course, where pollution is very heavy the distances involved may be enormous. Wiebe (1927) observed that the biological sequence following very heavy pollution extended 100 miles in the Mississippi, and Richardson (1921, 1929) watched the steady increase in the pollution of the Illinois river from 1913 to 1929 and the movement downstream of the various zones until they extended 250 miles from the source of the pollution near Chicago.

So far we have been considering only the general faunal composition of the zones and we find, as with the algae, that organic pollution causes a reduction in the number of species of invertebrates; a few species being, as it were, selected and presented with ample foodstuffs. This process leads to an increase in the total number of individuals in the animals as it does with the algae. The numbers of Tubificidae, *Chironomus* and *Asellus* are often enormous. In the Kalamazoo river in Michigan, Surber (1953) found 22 species, at a density of 729/sq. yd., above a sewage outfall: below the outfall only 3 species remained, mostly Tubificidae, but the numbers were increased to ten-fold, and even 30 miles below the outfall there was still a four-fold increase in the density of the animals although 16 species were present. A rather similar increase was observed in the Bristol Avon (Pentelow *et al.*, 1938), but in this river a further effect occurred.

In 1935 the river was receiving the effluent from a milk factory and, as shown in Table 9, this induced a five-fold increase in the faunal den-

TABLE 9

The numbers of animals per square metre at several stations in the Bristol Avon in the years 1935 *and* 1936, *during and after pollution by milk wastes. The numbers are based on four surveys carried out during the summers and autumns of the two years. Data from Pentelow et al.* (1938).

Position of station relative to outfall.	Average number of animals per sq. metre	
	1935	1936
2½ miles above	484	1,964
just above	1,336	3,528
300 yards below	5,788	19,940
0·7 miles below	2,076	24,864
2·2 miles below	2,352	6,460
5 miles below	764	3,232
7½ miles below	1,352	884

sity which persisted for 2 miles. In 1936, however, the pollution had ceased, but nevertheless the increase in numbers below the outfall was still present and it was very much more marked. The faunal composition at the various stations in the two years was more or less the same, although in 1936 some of the clean-water species, such as *Gammarus*, were relatively more abundant than they had been in 1935. Presumably the animals were, in 1936, living on the accumulated foodstuffs left by the previous pollution, and were able to take full advantage of them without the accompanying disadvantages of living in polluted water. Pentelow (Butcher *et al.*, 1937) has remarked that 'it seems that the effect of sewage pollution on the fauna of a river is a question of balance between unfavourable conditions such as low concentrations of oxygen, etc., and the advantage of increased food supply'. Here we have an example of the temporary existence of the advantages without the disadvantages.

Fishes

Fishes are usually eliminated for long distances by severe organic pollution, but the reason why this is so is not always clear. Obviously toxic substances, particularly ammonia, sulphides and cyanides, kill them, as do very low oxgyen tensions, and, as we have seen, low oxygen concentrations enhance the toxicity of most poisons. But fishes often disappear without apparently being killed. They are repelled by low oxygen concentrations, particularly at high temperatures (Jones, 1952), and if they are prevented from leaving relatively de-oxygenated water they develop clear signs of distress. For example their rate of gill movement increases directly as the oxygen content falls (Lindroth, 1949), and obviously a fish which is, as it were, panting for breath is in a less satisfactory state than one which is able to breathe normally. Several investigators have observed, however, that fishes were absent below the outfalls of organic effluents even though the oxygen content of the water always remained well above their lower limit of tolerance (Rasmussen, 1955; Pentelow *et al.*, 1938). At first sight it would seen that organic matter is repellent to them, and this may to some extent be true. But that it is not a complete explanation is shown by the well-known fact that fishes often congregate round sewage-works outfalls and apparently thrive on the rich food supply which is available there, in the form of worms and insects from the filters and particles of sewage fungus.

It has often been noted, however, that they tend to leave such places in warm weather. Perhaps this indicates that organic matter becomes repellent only when it is actively decaying, and support is lent to this idea by some recent work (Alabaster *et al.*, 1957) which has demonstrated that the concentration of carbon dioxide in the water has a great influence on the oxygen requirements of fishes. Carbon dioxide is produced in large amounts by the breakdown of organic matter, and concentrations of the order of 50 mg./l. may occur in polluted water. Such amounts have been shown to double the minimum concentrations necessary for the survival of fishes, and doubtless also they have similar effects on their avoiding reactions. This property of carbon dioxide may also explain anomalous results which have been obtained with caged fishes (Allan, 1955).

In a recent detailed study of the effect of a sewage effluent on fishes (Allan *et al.*, 1958) it has been shown that there is a very complex relationship between the various constituents. Thus, quite small amounts of carbon dioxide increase the susceptibility of fishes to lack of oxygen, which itself increases the toxicity of unionised ammonia. The toxicity of ammonia is, however, much reduced by the presence of carbon dioxide, which, by lowering the pH, causes the formation of the much less toxic ammonium ion. It therefore gives protection against ammonia poisoning up to values of about 30 p.p.m. of carbon dioxide, at which concentration this gas itself begins to be an important toxin. This study also indicates that detergents and metals in solution, even when present in amounts insufficient to kill fishes, increase their susceptibility to lack of oxygen. Undoubtedly therefore the dissolved oxygen content of the water is the factor of central importance, but its effect can be modified by a number of things, included amongst which is that most variable of influences, the weather. It is also probable that, under many circumstances, a combination of conditions which is toxic to fishes is also repellent to them.

We have also seen that some poisons, especially nitrites, act very slowly and so may eliminate fishes from reaches below outfalls as may substances like detergents. The latter are certainly toxic, although, unexpectedly and in a manner as yet unexplained, they become less so after passage through a biological treatment plant (Herbert *et al.*, 1957). It is possible that they are also repellent under certain conditions and the same may apply to other substances in organic effluents.

There are therefore several possible explanations for the absence of fishes from organically polluted water, and in our present state of knowledge we cannot choose between them except in particular instances. As one passes downstream fishes begin to reappear, usually in the lower parts of the *Cladophora/Asellus* zone, and coarse fishes reappear before trout. Very often the species which extends furthest upstream is the three-spined stickleback, followed by such species as roach, chub, gudgeon, tench, bream and perch, but much depends on the type of river involved, and on the degree of pollution. For instance very slight pollution, even if it induces the growth of sewage fungus for a short distance, may not apparently have any obvious influence on the fish population.

There is, however, one point in the life history of the fishes of eroding substrata at which even slight pollution may cause damage. As we already know, salmon and trout spawn in gravel, and their spawning sites may be seriously affected by quite small amounts of suspended matter. The same applies to sewage fungus, and it has been shown experimentally that trout eggs do not survive beneath a carpet of sewage fungus even if the water remains well oxygenated (Rasmussen, 1955). It is clear then that even if the fishes themselves survive in slightly polluted reaches they may suffer from some restriction of the area available for spawning. Here again much depends on the actual configuration of the river, and unless the spawning grounds are severely limited the results may be negligible as all fishes lay large numbers of eggs. Another aspect of fish ecology which should not be overlooked is that trout and young salmon, and indeed many of the coarse fishes also, hunt by sight; the Tubificidae and *Chironomus* of polluted water are therefore not available to them as food because of their burrowing habits. Slight organic pollution may therefore actually reduce the amount of food available to fishes by eliminating some of the normal fauna, even though it actually increases the numbers of invertebrates per unit area of river bed.

Chapter X

FURTHER BIOLOGICAL ASPECTS OF ORGANIC POLLUTION

IN the last chapter we considered the biological consequences of severe or moderately severe pollution of rivers by organic substances. There remain a few aspects of this type of pollution which we have still to discuss.

Very mild pollution

When organic pollution is very mild, because of efficient biological treatment or great dilution, it produces no obvious effects such as growths of sewage fungus or a significant fall in the oxygen content of the water. Although such phenomena may be detectable by detailed and careful study near to the outfall, they disappear after a very short distance. But mild pollution has biological consequences which can be detected for much greater distances.

In Chapter III it was pointed out that the algal flora of rivers changes as their fertility increases, and that in eutrophic waters the community which grows on slides comes to be dominated by *Cocconeis*, *Ulvella* and *Chamaesiphon*. This change from an oligotrophic to an eutrophic community may be brought about by enrichment of river water by the nutrient salts in an organic effluent. In the upper reaches of the river Tees, where the water is poor in nutrients, Butcher (1948) found that the normal oligotrophic algal community was replaced by the eutrophic community below several small sewage works, and that although the alteration persisted for only a short distance it occurred where there were no gross pollutional effects. The sewage was therefore acting as a manure, the consequences of which persisted only so far downstream as the concentration of nutrients remained higher than elsewhere in that part of the river. As soon as they had been used up, or diluted to a level normal to oligotrophic waters, the algal community reverted to its original condition; but this occurred much further downstream than it was possible to observe any of the more normal parameters of pollution.

In a rather similar manner it is possible to detect very slight organic pollution by study of the invertebrates of the river bed. Pentelow was the first investigator to report on this phenomenon (Butcher *et al.*, 1937), and some of his original data are reproduced here in simplified form (Table 10). He found that the invertebrate faunas of the river Tees above and below the outfall from Barnard Castle sewage works differed slightly,

TABLE 10

The percentage composition of the fauna of the river Tees above and below Barnard Castle sewage works. * = *less than 0·5 per cent. Data from Butcher et al.* (1937), *somewhat simplified in that percentages are given to the nearest whole number and species which made up less than 0·5 per cent. of the fauna at both stations are not included.*

	1½ miles above	300 yards below
Mayflies		
Baetis	4	3
Ephemerella	2	3
Ecdyonurus	13	5
Rhithrogena	24	3
Stoneflies		
Amphinemura	1	1
Leuctra	13	4
Other Plecoptera	2	1
Caddisflies		
Polycentropus	8	6
Hydropsyche	*	1
Agapetus	5	*
Other Trichoptera	3	1
Beetles		
Coleoptera	1	*
Non-biting midges		
Tanytarsus	3	33
Other chironomids	16	30
Limpet		
Ancylastrum	5	7
Average no. per square metre	136	847

and he attributed this to the effect of the sewage. As can be seen from the table, the total numbers of animals per unit area were increased about six-fold below the outfall, but the composition of the fauna was little altered. Pentelow remarks that 'the difference was qualitatively very slight', and he did not pursue this aspect of the matter further. However, in the light of my own research some of the changes in importance of the various species which he recorded seem very significant: these are the decreases in *Ecdyonurus*, *Rhithrogena*, *Leuctra* and *Agapetus* and the increases in *Tanytarsus* and other chironomid larvae (*Figs.* 5, 6, 7 and 8). The identities of the species in the latter group were not established, but most of those involved were placed in a category the members of which were described as 'small green forms'.

The increase in the density of the fauna was presumably attributable directly to the general enrichment of the environment and its effect on the algae. Although this does not seem to be a universal phenomenon we have seen, in Chapter IX, that similar increases in faunal density can occur after the cessation of severe organic pollution, probably for the same reason. The changes in faunal composition are, I believe, also explicable. Thienemann (1954) states that several members of the Orthocladiinae, the larvae of which are small and green, are favoured by pollution, and it would seem that in the Tees we have an example of this, and that the cause is probably a change in the type of algal, or detrital, food available to the larvae.

The changes in other members of the fauna were almost certainly due to suspended solids and the resulting slight increase in the siltiness of the environment. The two mayflies which were adversely affected were the two flattened forms which depend on clean stones for the efficient functioning of their hold-fast mechanisms. The stonefly, *Leuctra*, is a genus which is particularly dependent on being able to creep down amongst the gravel. The caddisfly, *Agapetus*, was the only species present which lives an exposed life on the stones, and it, like the flat mayflies, needs a clean firm surface on which to attach itself. It will be appreciated therefore that a slight increase in siltiness would reduce the amount of suitable living space for these creatures and so lead to a decline in their importance. *Tanytarsus*, on the other hand, showed a marked increase in numbers. We already know that the larvae of this genus build their tubes of silt and are often very abundant in the *Cladophora* zone; it is therefore to be expected that they would be favoured by an increase in the amount of silt in the river bed. It can be seen then that the observed changes can be explained quite simply in terms of increased fertility and siltiness.

Another, more complex, example of mild organic pollution is illustrated by Table 11. Here two sets of samples are shown, taken from the Welsh Dee in two different years. Once again this table illustrates the differences in quantitative results which may be obtained by the same technique at different river levels. In January 1956 the water level

TABLE 11

Numbers of animals collected by a standardised netting technique, and estimates of the abundance of plants, at various points in the Welsh Dee. Only selected organisms are shown. There are two sets of samples collected in January 1956 and April 1957, and these are distinguished by different types (ordinary = 1956, bold = 1957) (P = present, C = common, A = abundant). An organic effluent entered the river above station D and a second, minor, effluent entered above station C.

Station No.:	A	B	C	D	E	F	G	H	I	J	K	L
Distance above outfall Distance below outfall (miles)	1·3 —	0·4 —	0·2 —	— 0·2	— 0·6	— 1·1	— 2·6	— 4·7	— 5·9	— 7·8	— 10·2	— 16·0
Flatworms *Polycelis felina* .	1 **21**	2 **1**	5 **3**	— **—**	— **1**	— **—**	— **—**	— **—**	— **—**	— **—**	— **—**	— **—**
Worms Lumbriculidae .	34 **8**	6 **7**	52 **6**	— **—**	6 **2**	13 **1**	15 **55**	7 **72**	23 **127**	17 **163**	10 **24**	130 **437**
Shrimps Amphipoda .	6 **20**	22 **—**	32 **3**	— **—**	2 **2**	8 **6**	4 **—**	1 **1**	4 **2**	10 **2**	9 **15**	— **6**
Mayflies *Baetis* . .	19 **337**	50 **540**	40 **205**	1 **36**	10 **385**	4 **205**	1 **716**	1 **428**	2 **502**	3 **281**	— **500**	1 **314**
Ecdyonurus venosus . .	12 **23**	12 **32**	— **5**	1 **4**	— **15**	— **26**	— **1**	1 **14**	2 **8**	4 **4**	14 **45**	4 **18**
Rhithrogena semicolorata .	2 **14**	— **41**	— **13**	1 **—**	— **9**	— **33**	— **23**	— **56**	9 **241**	1 **123**	— **126**	1 **69**
Heptagenia sulphurea .	4 **29**	29 **24**	7 **8**	— **2**	— **5**	— **4**	— **3**	— **4**	— **1**	— **3**	1 **74**	— **17**
Stoneflies Plecoptera . .	13 **40**	16 **55**	5 **16**	1 **5**	— **42**	— **26**	— **12**	— **21**	5 **82**	5 **71**	3 **127**	6 **42**
Caddisflies Trichoptera .	13 **26**	43 **60**	19 **30**	1 **18**	4 **43**	5 **23**	— **36**	9 **36**	6 **61**	23 **62**	10 **112**	7 **54**
Beetles Helmidae . .	21 **53**	8 **26**	20 **8**	— **4**	1 **31**	— **11**	1 **23**	5 **24**	11 **96**	25 **91**	13 **87**	261 **330**
Biting midges Ceratopogonidae .	— **21**	— **21**	2 **10**	— **151**	— **123**	— **96**	1 **153**	— **76**	— **67**	— **23**	— **56**	— **18**
Non-biting midges Chironomidae .	17 **63**	38 **104**	43 **54**	9 **67**	17 **89**	96 **28**	18 **115**	6 **52**	9 **33**	13 **26**	25 **86**	11 **74**
Limpets *Ancylsatrum fluviatile* . .	3 **1**	4 **1**	3 **3**	— **—**	1 **—**	— **—**	— **—**	— **1**	1 **13**	2 **9**	— **51**	21 **87**
Algae *Lemanea* . .	C **A**	C **A**	C **C**	— **—**	— **—**	— **—**	— **—**	— **—**	— **—**	— **—**	— **—**	— **—**
River mosses *Fontinalis antipyretica* .	C **A**	A **C**	A **A**	— **P**	— **C**	— **C**	— **C**	— **—**	— **C**	— **—**	P **—**	— **C**
Water crowfoot *Ranunculus* .	C **C**	C **C**	C **C**	— **—**	— **C**	— **C**	— **C**	— **C**	— **—**	— **A**	— **A**	— **A**
Total no. of animals in sample .	304 **756**	263 **1,037**	303 **453**	66 **304**	53 **789**	133 **480**	44 **1,161**	39 **818**	90 **1,250**	135 **889**	88 **1,373**	469 **1,495**

was normal, but in April 1957 it was very low; the last line of the table shows that many more animals were caught in 1957. Otherwise, however, the two sets of samples are comparable, and they were taken at biologically similar seasons, i.e. late enough for the spring insects to be large enough to catch and early enough for no appreciable losses to have occurred through emergence.

A small amount of organic matter entered the river between stations B and C, from a rural sewage works and a dirty brook, and these sources of pollution were the same in the two years. Between stations C and D the river received a much heavier load of organic matter from an industrial process, but the quality of the effluent, which was biologically treated, was much better in 1957 than it had been in 1956.

Considering first the more upstream source of pollution, and comparing the faunas of stations, A, B and C in both years, it will be seen that there was virtually no change in the *composition* of the fauna at station C, but a few creatures became less common. These were once again, most notably, the flat mayflies, *Ecdyonurus*, *Rhithrogena* and *Heptagenia*, but there was also a considerable drop in the importance of *Baetis*, stoneflies and caddis-worms. This then is just the effect observed by Pentelow in the Tees below Barnard Castle, but here there was no increase in numbers. Indeed in 1957 there was a decrease, nor did any animal increase markedly in importance, with the possible exception of Lumbriculidae in 1956; these worms, however, tend to fluctuate rather unexpectedly in abundance, see for example stations J, K and L in 1957. Very mild organic pollution thus seems to act adversely on Ecdyonuridae, *Baetis* and stoneflies, but without totally eliminating them.

One can indeed go further, although the following details are omitted from the table in order to keep it a reasonable size. Within the genus *Baetis* the common species *rhodani* seems to be the more able to withstand mild organic pollution than are the other species. Among the stoneflies *Amphinemura sulcicollis* (*Fig.* 6) seems to be the least affected and it is perhaps significant that this species is more tolerant of silt then are most other Plecoptera (Hynes, 1941). In contrast the spring species of *Leuctra* are among the first animals to be affected, but the summer species *L. fusca* is almost as tolerant as *A. sulcicollis*. Probably this is because *L. fusca* is one of the very few stoneflies which are adapted to withstand the relatively low oxygen concentrations and periods of low water, with its attendant silting, of the summer months: as was stated in Chapter IV most stoneflies avoid the warm summer months altogether. It will be apparent, therefore, that in the study of very mild pollution identification to species is as important as it is in the Protozoa inhabiting severely polluted waters.

Returning now to the river Dee and Table 11. The effluent entering below station C was only mildly de-oxygenating, and this effect could be detected only at station D and even there was hardly apparent in 1957. It had, however, the property of stimulating the growth of *Sphaerotilus*. In 1956 there was a heavy growth of this bacterium at station D, and scattered sparse growths extended as far downstream as station L. In 1957, however, when the effluent was receiving efficient biological treatment, *Sphaerotilus* could be found only at station D, and even there it was very sparse and patchy. We are able, therefore, to compare the effects of two degrees of mild pollution by the same effluent.

In 1956 the effluent caused a general reduction in the fauna and the total elimination of the larger plants for a long distance. Many animals were absent at station D, and others were absent for long distances downstream. Once again the Ecdyonuridae and stoneflies were the most seriously affected, but the Helmid beetles and the limpets were also absent for a considerable distance, and the fauna was only approaching its normal condition 6–8 miles downstream of the outfall. Even there, however, flatworms were still absent and mayflies and limpets were scarce.

In 1957 the effect of the effluent was almost confined to station D, but a few organisms were more seriously affected. Thus *Lemanea* and *Polycelis felina* (*Fig.* 5) were eliminated for at least 16 miles, and limpets were absent for 5 miles. So even though, in general, the effect was only observable for half a mile or so the effluent was biologically detectable for a long distance. This becomes even more apparent when the data are considered in greater detail than is shown in the table. Taking a few taxonomic groups and considering them in turn, we find the following facts:

1. Leeches (not shown on the table). Two species of the genus *Erpobdella* were present above the outfall: both were absent from station D, but *E. octoculata* (*Fig.* 5) returned at station F. In 1956 only *E. testacea* was found below the outfall, and we have already seen that the latter particularly favours organically polluted water.

2. Shrimps. *Crangonyx pseudogracilis* returned at station E, but *Gammarus pulex* (*Fig.* 6) was absent above station I, although present below this point and above the outfall. In 1956 only a single *G. pulex* was found below the outfall at station I.

3. *Baetis*. *B. pumilus* was absent from station C and downstream as far as station H. *B. rhodani* (*Fig.* 5), on the other hand, was present all the way. Clearly then even the very mild pollution above station C was too much for the former species.

4. Ecdyonuridae. As can be seen from the table *Heptagenia* was more

seriously affected in both years than were *Ecdyonurus* and *Rhithrogena*.

5. Stoneflies. Four species were fairly common in the river and of these *Isoperla grammatica* (*Fig.* 6) alone was present at station D. *Amphinemura sulcicollis* reappeared at E, *Chloroperla torrentium* at G, and *Leuctra inermis* at I.

6. Caddis-worms (*Fig.* 7). *Polycentropus*, *Hydropsyche* and *Rhyacophila* were present at station D. *Sericostoma* reappeared at station E, but *Glossosoma*, which resembles *Agapetus* in appearance and habit, reappeared only at station F.

7. Helmidae. Four species of these little beetles were present. Two, *Esolus parallelopipedus* and *Limnius tuberculatus*, were present at station D; *Latelmis volkmari* reappeared at E and *Helmis maugei* (*Fig.* 6) at G.

During 1957 the quality of the effluent was further raised, and a similar set of samples collected in March 1958 showed still further improvement. The detailed findings are not included in the table, in order not to make it too long, but the relevant points were: *Sphaerotilus* had disappeared entirely except just at the effluent outfall; *Lemanea* had recolonised the river except for the first 2½ miles; *Erpobdella octoculata*, *Crangonyx*, *Glossosoma* and *Ranunculus* had reappeared at station D, and *Gammarus* and *Helmis* at station E. There was also no longer any sharp fall in numbers of *Baetis* and caddis-worms, nor in the total numbers of animals, caught at station D. Many sensitive creatures had therefore moved upstream and others no longer showed any effect of the effluent. The only effects of the previous pollution which persisted for any distance were the continued absence of *Polycelis* for at least 5 miles, the absence of limpets for 4½ and of *Lemanea* for 2½ miles.

Perhaps to the reader this seems a rather dry catalogue of data, but it is included here in order to emphasise that, at this very slight level of pollution, specific differences are important, and that they enable one to detect an effluent a considerable distance below the reaches where it exerts any clear effects. In this instance fishes were apparently quite unaffected, even in 1956 at station D, and chemical investigation revealed very little change in the water; but careful biological study of the river showed alterations in the fauna and flora extending for at least 16 miles, and the steady decline of these effects as the effluent was improved. This example also illustrates how an industry faced with a pollution problem can, by installation of suitable treatment plant and careful control of the quality of its effluent, return a river to its normal biological condition while continuing to use if for waste disposal.

It is difficult in our present state of knowledge to summarise or explain the biological consequences of very mild organic pollution, but it should be clear from the above examples that there are a few general principles. Some flatworm species may be eliminated; flat mayfly nymphs are severely affected, and *Baetis rhodani* seems to be a resistant mayfly. Stoneflies vary in their susceptibility, *Isoperla* and *Amphinemura* being more resistant than *Leuctra*. Case-bearing caddis-worms are more affected than caseless species, and among the former those, such as *Agapetus* and *Glossosoma*, which live exposed lives on the tops of stones are less able to withstand pollution than others. Among the Helmidae the different species vary in their susceptibility. Clearly one cannot be dogmatic about these points until much more work has been done on the subject, but they give some indication of the sort of information which can be obtained.

Recovery from pollution

When organic pollution suddenly ceases, recovery may be very rapid if there are suitable reaches from which normal river fauna can recolonise the polluted area. Fishes, as we know, move in very rapidly, but invertebrates and plants are much less mobile; they can nevertheless reappear after quite short periods. Pentelow (Butcher *et al.*, 1931) observed that the fauna of the river Lark was reduced every winter to Tubificidae by pollution from a sugar factory, and he showed that *Sialis*, *Gammarus*, leeches and snails were actually killed in the river by the effluent. Nevertheless, during the spring and summer these creatures reappeared, having presumably moved in from tributaries. This must have been due to immigration and not to reproduction, because the interval was too short to have allowed even one generation to occur, and it was possible to explain the differences in rate of recolonisation at the different stations in terms of the nearness of suitable tributaries from which immigration could occur.

I have also observed rapid recolonisation of a small stony stream in southern England. This had been grossly polluted by sewage, and the fauna reduced, presumably, to Tubificidae, although it was not examined when it was in this condition. The pollution continued for some years, but one September it suddenly ceased when new arrangements for disposal of the effluent were made. In the following April the fauna was found to be rich and varied, including flatworms, leeches, *Asellus*, *Gammarus*, *Baetis*, *Polycentropus*, *Stenophylax*, Chironomidae and *Limnaea pereger* (*Figs*. 5, 6, 7 and 8). These can only have come from several tributaries which entered the stream along its length, because the period September to April was not long enough for any of these creatures to have bred and grown to a reasonable size. The tributaries, however, also contained other animals which had not yet appeared in the main stream. These included *Rhithrogena*, *Ecdyonurus*, *Leuctra*, *Nemoura*,

Amphinemura, *Agapetus* and *Ancylastrum* (*Figs*. 5, 6 and 7). If one compares the lists of the species which had and which had not recolonised the main stream during the six winter months it is at once apparent that the former is composed of animals which are mobile and fairly readily swept away by the current. The latter list consists of species with good hold-fast mechanisms (flat mayflies, *Agapetus* and limpets) or cryptic habits (the stoneflies).

The implications are therefore that invertebrates tend to recolonise formerly polluted reaches in succession, and that those which are not easily moved by the current are the last to reappear. But this applies, of course, for only a short period; as soon as spring comes the insects are able to disperse readily as flying adults and lay their eggs in the affected reaches. Thus, after even a single breeding season, all the insects can be expected to reappear. The molluscs, however, have no such dispersal mechanism and have to crawl back, and this may take them a long time. For this reason it is probable that the absence in 1957 and 1958 of limpets from several stations in our second example of mild organic pollution (Table 11 and page 124) was due not to the actual effect of the effluent in those years but to a delayed effect. They simply had not had time to recolonise a long reach of the river; but by the spring of 1959 they were present at all stations.

A similar instance of differences between the rates of recolonisation was observed in the river Lee. This was, as we have seen, polluted by sewage, but by 1952 there were marked signs of recovery and a rich fauna had been built up. This consisted mainly of *Asellus*, caddis-worms, *Haliplus* (*Fig*. 12), Chironomidae of various types and the snail *Limnaea pereger*, and included *Gammarus pulex* and *Baetis rhodani*. In August 1953, however, because of an industrial accident a large amount of copper cyanide entered the river through the sewage works. The effects of this on the fauna were serious but, as is typical of poisonous pollution, selective. The caddis-worms, *Haliplus*, Orthocladiinae, Tanypodinae and *Chironomus thummi*, were apparently quite unaffected, but the other creatures were eliminated for long distances. *Asellus* disappeared for 3 miles, *Gammarus*, *Baetis* and *Tanytarsus* (*Fig*. 10) for 8 and *Limnaea* for 16 miles. There were few sources from which recolonisation could occur between the outfall and mile 10, but despite this almost all these species had regained their former status in the river by the autumn of 1954. *Limnaea pereger*, however, which, unlike the crustaceans and the insects, is unable to fly or to be swept easily along by the current, had moved downstream only half a mile and upstream only to mile 8½.

It will be appreciated therefore that shortly after its cessation any kind of pollution can be detected by the absence of some kinds of insects, and that even a year later some slow-moving animals, particularly molluscs, may be absent from or scarce in the affected reaches unless there are very readily available sources from which recolonisation can occur.

Estuaries

In this book we are concerned primarily with pollution of inland waters, but because of their influence on migratory fishes we must briefly discuss estuaries. Many of these are the sites of large towns and industries, and in this country, and indeed in most others, many effluents receive no treatment at all if they are discharged into tidal waters, there being an erroneous but comfortable belief that there is always adequate dilution. The result is that many estuaries are very foul (Pentelow, 1955), so much so that Manhattan has been defined as a 'body of land entirely surrounded by sewage' (Phelps, 1944). Similar definitions could be applied to islands in the Thames estuary.

The discharge of large volumes of untreated sewage into an estuary results of course in severe de-oxygenation, and unless it is very large, or, like the Mersey, subject to a good deal of tidal ebb and flow (Department of Scientific and Industrial Research, 1938), this may be complete in warm weather. The result is that salmon and sea trout are unable to pass in either direction, and as these fishes tend to return to spawn in the river of their birth the local population soon becomes extinct. It is common knowledge that the Thames was once a first-class salmon river and now contains none of these fishes, and the estuaries of several other British rivers are in almost as bad a condition (Pentelow, 1955). It is not known if eels are similarly prevented from entering rivers through polluted estuaries, but it would seem very probable that they are and that small elvers are unable to swim any considerable distance through de-oxygenated water. Eels are still present in the Thames, where large specimens are occasionally caught, but these fishes can travel along the canals which link up most of the major English watersheds.

Apart from the effect on migratory fishes there is little detailed information on the biological consequences of estuarine pollution. Only two estuaries have been intensively studied with special reference to pollution, and one of these investigations, that on the Mersey (Department of Scientific and Industrial Research, 1938), was concerned primarily with physical and chemical studies, with the object only of ascertaining if the discharge of raw sewage was causing the deposition of mud. This proved not to be so. The second estuary, that of the Tees, was also studied biologically (Alexander *et al.*, 1935), but without discovering any very clear-cut biological consequences of the pollution. This is primarily because of the complexity of estuarine conditions. As is well known, estuaries are subject to abrupt changes in salinity caused by the ebb and flow of the tide and the fact that fresh water flows over the top of salt water, which is more dense. Mixing is therefore only partial, salinity varies with depth and the salinity changes at any given point are complex and often abrupt. Generally speaking freshwater creatures cannot stand much salt and marine creatures cannot stand much freshening of the water, and, although there are a few species which are adapted to brackish conditions, most plants and animals find the sudden changes in estuaries intolerable.

There is therefore always a sharp decline in the numbers of species which occur in the reaches subject to great salinity changes, although the animals which can tolerate the conditions are often abundant. Among these the following are often particularly numerous: the worms *Clitellio arenarius*, *Arenicola marina* and *Nereis diversicolor*, the crustaceans *Corophium volutator*, *Crangon vulgaris* (the edible shrimp) and *Carcinus maenas* (the shore crab), the snail *Hydrobia ulvae* and the bivalves *Macoma balthica*, *Cardium edule* (the cockle) and *Mya arenaria* (the clam), the winkle *Littorina littorea*, and among fishes the flounder. Presumably some, or possibly all, of these species are affected by pollution, but in the Tees the investigators were unable to distinguish changes due to pollution from those due to differences in substratum. The only clear difference they found between the Tees and the unpolluted estuaries of the Tay and the Tamar was that, in the absence of pollution, there were more species at the seaward end of the estuary. The Tees estuary was, however, at the time it was investigated not very badly polluted, and salmon were still able to pass through it: although the oxygen content was sometimes reduced to low levels it was never reduced to zero. Investigation of a grossly polluted estuary would probably reveal that some of the normal estuarine animals were absent. Similarly, investigation of local pollution in otherwise clean estuaries might furnish evidence of its effect on estuarine animals. For instance Stopford (1951) observed that the two burrowing Polychaete worms, *Pygospio elegans* and *Eteone longa*, which are widespread in the estuary of the Welsh Dee, were absent from the polluted area near a small town, and she quotes the deaths of lugworms and cockles during a warm summer as having been attributed to the enhanced effects of pollution.

There is virtually no information on the effects of pollution on the smaller inhabitants of estuaries, but one very characteristic feature of all organic pollution of marine waters is the development of *Beggiatoa alba*. This replaces the usual freshwater sewage fungus *Sphaerotilus* which cannot tolerate salt, and it forms a paper-white crust on the mud near drain outfalls and is sometimes accompanied by purple sulphur bacteria (Liebmann, 1951). Beneath the crust the mud is always black with ferrous sulphide and smells strongly of sulphuretted hydrogen. This blackening of marine muds is, however, a common feature, because sulphate is abundant in sea water and is readily reduced to sulphides by local decay. *Beggiatoa*, which is a sulphur bacterium, therefore occurs very widely on the seashore and extensive growths may occur quite naturally in stagnant inlets.

Hygiene

The last aspect of organic pollution which we have to consider is that of hygiene. This is, of course, primarily a medical subject and can be reviewed only briefly here; it has been discussed at length by Phelps (1944) and more briefly by Hawkes (1957). It is a well-known fact that many human diseases can be carried by water; they include typhoid fever, cholera, bacillary and amoebic dysentery, various parasitic worms and very probably also some virus diseases including poliomyelitis. The causative organisms of all these diseases have been detected in sewage.

Ordinary biological treatment is designed to oxidise organic matter and is not primarily concerned with the removal of pathogenic organisms. Although sewage works of ordinary design remove 50–99 per cent. of all bacteria (Southgate, 1951) they do so quite incidentally, and even if the effluents are then run through slow sand-filters a large number of bacteria remain, including of course some of the pathogens. In all probability viruses are little reduced by any form of treatment, and quite large proportions of the eggs and cysts of parasitic animals pass through ordinary sewage works (Wang and Dunlop, 1954) and even through rapid sand-filters (Silverman and Griffiths, 1955). Water contaminated by sewage therefore presents a definite risk of infection to bathers and those who drink it, and this risk is only reduced by ordinary biological treatment.

We know very little about the fate of virus particles in water, but it is well established that most of the pathogenic bacteria are unable to reproduce outside the body and that they slowly die off in river water, each type having a characteristic death-rate which becomes more rapid the higher the temperature. Thus the further downstream from an outfall the lower becomes the concentration of pathogenic bacteria, but it has been suggested that the greater the pollution the less rapidly this occurs. Liepolt (1958) stresses the fact that if conditions are suitable for the growth of sewage fungus they may ensure the better survival of pathogenic bacteria. But bacterial self-purification does occur, even in polluted water, and it is probable that the same applies to the eggs of intestinal worms, such as *Ascaris* (the roundworm) *Trichostrongylus*, *Trichiuris* and *Taenia* (the tapeworm) and the cysts of *Entamoeba*, one species of which causes dysentery. Resting stages of all these parasites have been recovered from sewage, and Silverman and Griffiths (1955) have reviewed the literature on the dispersal of worm eggs and protozoan cysts in this medium. Self-purification is, however, clearly of far less significance for these animal resting stages than it is for bacteria which are much less viable. The eggs of the beef tapeworm, *Taenia saginata*, have been shown to survive for many months in sewage sludge (Silverman, 1955), and more than 80 per cent. of the roundworm eggs recovered from sewage were found to be alive (Wang and Dunlop, 1954). The tapeworm is, of course, not directly infective to man and has to be eaten by cattle before becoming so, but beef-cattle often have access to polluted water.

Silverman calculated that in the sewage he was investigating there were between 1 and 6 eggs per gallon, and this, coupled with the preference of many people for steaks which are almost raw, could represent a considerable risk of infection. Even the most thorough and conscientious meat-inspection would be unlikely to detect very light infections. Moreover, Silverman showed that the eggs persisted in sludge which was to be used as manure, and that birds, particularly seagulls, feeding on sewage works could disseminate them, as they pass undamaged through their intestines. Similarly it is well established that the human roundworm can be spread to land which is fertilised with sewage sludge (Silverman and Griffiths, 1955). It is thus likely that the eggs of parasitic worms are readily and widely dispersed from a sewage works, either directly in sludge or indirectly in the effluent or in detergent foam carried by the wind from activated-sludge tanks.

It would at first sight seem an obvious solution of the problem if all sewage effluent were to be chlorinated as are most water supplies. This would at least eliminate the causative organisms of bacterial and virus diseases, although it is doubtful if it would affect worm eggs. But from other points of view chlorination would be highly undesirable. It would sterilise the effluents and so greatly delay the process of self-purification in the rivers. It would also be dangerous to the normal inhabitants of the rivers because chlorine is very toxic, as little as 0·3 mg./l. being lethal to fish (Allen *et al.*, 1946). Although organic matter combines with and inactivates free chlorine, the 'chlorine demand' of sewage varies widely and irregularly (Southgate, 1951); it would thus be virtually impossible to regulate the dosage so that the bacteria were killed and no free chlorine remained in the effluent. Then again there is the fact, which we have already discussed, that chlorine forms very toxic compounds with some common constituents of city sewage. The general approach has therefore been to attempt to assess the risk and to define standards to which at least drinking water, public swimming baths and shellfish must conform.

Unfortunately detection of pathogenic bacteria is not easy and so it has become the practice to assess their presence at second hand, rather in the same way that B.O.D. is used in the assessment of organic matter. *Escherischia* (*Bacterium*) *coli* is a common non-pathogenic intestinal bacterium which can be detected by a fairly simple technique and which slowly dies off outside the body but more slowly than do the pathogens. It therefore provides a convenient measure of the sanitary state of water. It can, however, reproduce in water containing organic matter (Allen *et al.*, 1952), especially in the *Sphaerotilus* zone (Bahr, 1953), and for this reason it has been suggested that another non-pathogenic intestinal bacterium, *Streptococcus faecalis*, might be a more suitable organism to use (Mallman, 1940). This idea has, however, not been widely accepted, and indeed it would seem preferable in this instance to use an organism which was likely to give a margin of error which tends to make conditions appear worse than they are: it is better to err on the side of safety. The sanitary condition of water is therefore generally assessed in terms of the density of *E. coli*, often supplemented by so-called plate-counts in which the total numbers of bacteria growing on gelatin at 20° C. or on agar at 37° C. (body heat) are determined. The numbers are usually quoted as numbers per millilitre of the original sample.

Permissible standards are then fixed on the basis of such counts according to the particular use for which water is required, but it will be appreciated that they result only in a degree of safety. As long as some *E. coli* is present there is a chance that some pathogenic bacteria also remain; and these standards apply only to bacteria, with primary reference to typhoid; worm eggs are almost certainly, and some viruses probably, more resistant than *E. coli*. Moreover, no attempt is made to ensure that effluents entering rivers which are not used for water supply should conform to definite standards. The risk to public health from open waters therefore remains unchanged, and people disporting themselves in or on polluted waters take a definite, but usually slight, risk of contracting disease. This risk varies with the temperature, the distance from the outfall and the state of health of the population from which the sewage originated.

Chapter XI

HEAT, SALTS AND POLLUTION OF LAKES

Heated effluents

THE biological effects of the discharge of clean but hot effluents into rivers and streams depend on how much the temperature is raised. If the increase is small there is probably only a general speeding up of biological processes in the water, but if there is a steep rise of temperature, even of the order of 5–10° C., there may be clear biological consequences. If temperatures of about 40° C. are exceeded most normal river creatures will be eliminated, and they will reappear further downstream only as the temperature falls, in much the same way as they reappear below the outfalls of poisonous effluents.

We have very little information on the temperature tolerances of freshwater plants and animals and, as far as I am aware, no field-study has been made on a simple example of thermal pollution. We know, however, that fishes vary greatly in their ability to withstand heat, and that the upper limits for different species vary from 22° C. to more than 42° C. (Huntsman, 1942). Trout die at about 25° C. (Huntsman, 1942; Gardner, 1926), their eggs will not hatch at temperatures higher than 14·4° C., and they grow more rapidly at temperatures below 15·5° C. than they do at higher temperatures (Southgate, 1951); they are thus clearly cold-water fishes. On the other hand carp can withstand temperatures of up to 35–38° C., the lower figure being the limit for large individuals and the higher that for small ones (Meuvis and Heuts, 1957), and pike and goldfish occupy intermediate positions (Gardner, 1926).

These differences in heat tolerance probably account at least in part for the normal distribution of fishes in river systems. Apparently under normal circumstances fishes are able to avoid lethal temperatures by remaining in their appropriate zones, but there have been reports of naturally occurring deaths caused by high temperatures. Huntsman (1942) states that salmon are killed in hot summers in Nova Scotia at times of very low water, and that the largest specimens die first. He also found evidence of acclimatisation in that the fishes which had recently entered the rivers died when the temperature reached 29·5° C., while those which had been there some time survived until there had been a further rise of a degree. The smallest specimens, or parr, which had not yet been down to the sea, tolerated even higher temperatures.

We may conclude therefore that fish species vary in their temperature tolerance, that this may change with age and that there is also probably some amount of acclimatisation. Doubtless the same applies to river invertebrates, but on this point there is virtually no information. It would seem at first sight that the maximum, or summer, temperatures are the ones which matter most, but the fact that, for instance, trout eggs require low temperatures indicates that winter temperatures may also be important. It is probable, for example, that some insect species, particularly stoneflies which emerge very early in the year and pass the warmer months in the egg stage, actually require very low winter temperatures.

We can, however, take this subject no further at present because of lack of information. Suffice it to say that it is certain that a persistent temperature rise of the order of 10° C., would produce a clear effect on a trout stream by the elimination of some species, including, probably, the trout themselves, and that this effect could be achieved in one or both of two ways. Either summer temperatures would exceed the tolerated maxima of some species, or winter temperatures would be too high for certain stages in their life histories. According to Butcher (Rees, 1954) thermal pollution does not appear to affect plant communities, but obviously this statement is applicable only where the temperature rise is not very great. At least one plant, *Vallisneria spiralis*, which is grown in many warm aquaria, has become naturalised in Britain in places where water is heated by effluents; it would be unable to survive here if exposed to ordinary winter temperatures. The same seems to apply to some animals. Recently Mann (1958) has reported the occurrence of the tropical tubificid *Branchiura sowerbyi* in the Thames, at a point where the water is warmed by a power-station effluent: previously this species was known in this country only from hot-houses. Similarly Naylor (1958) in a paper to the XVth International Congress of Zoology recorded a number of exotic marine animals from an artificially warmed dock in Swansea. Two of the species, a crab and a barnacle, are only able to survive in Britain under the conditions produced by heated effluents.

Heated effluents are, however, not normally discharged into clean water, although this will undoubtedly become a more common practice when atomic power stations are more abundant, because most of them are to be sited in remote areas. At present the majority of heated effluents are produced by industries and conventional power stations, which are situated in densely populated regions, and the water which most of them use for cooling purposes is therefore already polluted. We have already noted the consequences of this practice in Chapters V and VI. To recapitulate briefly there are two of these. Firstly it increases the B.O.D., by killing the bacteria, small plants and animals in the water and turning them into dead organic matter. This results either from the heat itself or from chlorination designed to prevent growths, chiefly of sewage fungus and similar bacteria, in pipes. Secondly it may make the water toxic either with residual chlorine or with compounds such as cyanogen chloride. The final effluent is therefore more polluting than was the water at the intake.

Apart from these properties of hot effluents the heating of a polluted

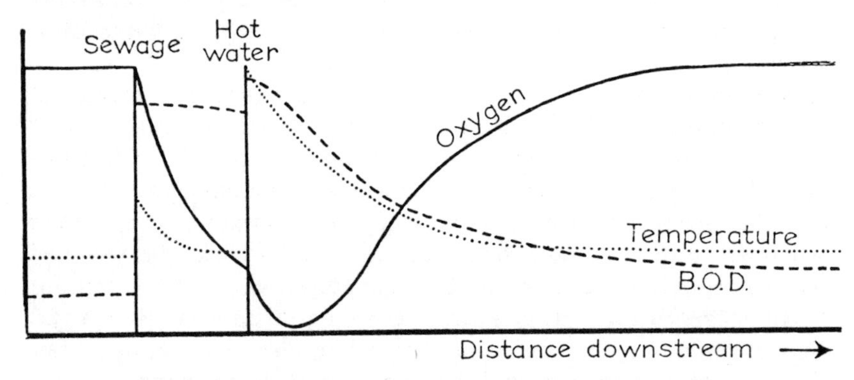

FIG. 22. Diagrammatic presentation of the effect of a heated effluent on an organically polluted river.

river has profound biological effects. It stimulates the growth of sewage fungus, for which the optimum temperature lies within the range 20–30° C. (Southgate, 1951), and it speeds up all other biological activity. This is illustrated diagrammatically in *Fig.* 22, which should be compared with section A of *Fig.* 16. In *Fig.* 22 the initial conditions are the same as those in *Fig.* 16*A*, but some distance below the sewage outfall the river receives cooling water from, say, a power station. It will be seen that this not only raises the B.O.D. in the way we have discussed already, but it increases the rate of oxidation of the organic matter and so deepens the oxygen sag. It therefore causes a marked deterioration of conditions, although at the same time shortening the various zones of pollution because the organic matter is destroyed more rapidly. It will, however, be appreciated that the beneficial effect is likely to be outweighed by the greater severity of the pollution further upstream. In the hypothetical example illustrated very little further increase in the pollution would result in total de-oxygenation and septic conditions.

Increase in temperature also increases the toxicity of such poisons as are present (Chapter VII), and the greater rate of oxidation results in higher production of carbon dioxide, which itself raises the lowest oxygen content which fishes can tolerate (Alabaster *et al.*, 1957): this last point applies probably to invertebrates also. Higher temperatures also increase the speed at which fishes react to low concentrations of oxygen (Jones, 1952) as well as directly increasing the minimum percentage saturation with oxygen necessary for their survival (Downing and Merkens, 1957). It will thus be seen that the heating of a polluted river, even without the usual increase in B.O.D. and the probable addition of extra poisons, enhances the effect of toxic and organic pollution. Although the higher temperature may enable fishes to escape from regions of low oxygen tension, rather than remaining to be killed by asphyxiation, it effectively clears them out of affected reaches. The more sensitive invertebrates, which might otherwise have survived, are removed and the river is impoverished generally. On the other hand the pollutional effects of organic matter, although more severe, may extend for a shorter distance. This is, however, unlikely to be of any consequence except in very mild cases because of the long distances needed for recovery and, of course, it does not apply to pollution by poisons unless they are chemically unstable.

Non-toxic salts

Pollution by non-toxic salts is produced in various parts of the world by mining operations and oil drilling, but it has been little studied in this country, where it is relatively rare. Most freshwater animals are unable to withstand very high concentrations of salt, and the actual composition of the solution is also important because a mixture is usually more tolerable than an equivalent concentration of one type of salt. As with poisons, which are effective at much lower concentrations, the salt tolerances of the various species of plants and animals differ. Thus, for example, the pondweeds *Potamogeton pusillus*, *P. filiformis* and *P pectinatus* are commonly found in brackish water, whereas other members of the genus, and indeed most freshwater weeds, do not tolerate much salt. Similarly the milfoil *Myriophyllum spicatum* occurs commonly in coastal saline waters, and this species and *P. pectinatus* are the major constituents of the flora of inland saline waters in the Midlands. Likewise it has been shown that animals vary widely in their tolerance to brines from North American oilfields. These brines, which are mainly chlorides, eliminate many crustaceans and insects, but fishes seem to be more tolerant (Lafleur, 1954). Some invertebrates, however, including crayfish and dragonflies, are able to withstand concentrations of salt of over 15,000 mg./l. (Clemens and Jones, 1954), and under some circumstances a special saline community may appear.

Liebmann (1951) lists a number of diatoms and a few animals, including the brine shrimp (*Artemia salina*) and the salt fly (*Ephydra riviparia*) which are characteristic of very saline water, and Claassen (1926) studied a small stream in New York State which was completely altered by salt pollution. This was a small tributary of the Genesee river, which ordinarily would have been a trout stream but whose water contained between 1 and 4 per cent. of sodium chloride. As a result the normal stream inhabitants had been replaced by salt-tolerant species. The bed was covered with the green alga, *Enteromorpha intestinalis*, a species which normally lives in brackish habitats and which is a familiar plant of freshwater seepages on the sea-shore. The animals included the salt fly *Ephydra*, the larvae of biting midges (*Culicoides*), mosquitoes, rat-tailed maggots, and a few worms and beetle larvae. Such drastic changes have not been reported from Europe, and in Germany many species of fishes inhabit rivers which have a fairly high salt content, to which they appear to be more resistant than many species of invertebrates (Schmitz, 1958); *Gammarus pulex*, for example, can tolerate very little salt.

Freshwater animals which, in my experience, thrive in fairly saline British waters include Tubificidae, *Asellus aquaticus*, dragonflies, some species of caddis-worm, *Chironomus thummi*, several species of water boatmen and water beetles, and some snails, most notably *Limnaea pereger* and the little operculate *Hydrobia jenkinsi* (*Figs*. 7, 10 and 11). It would seem that some species of flatworms, and all mayflies and stoneflies are adversely affected by salt, and that some fishes, especially roach, can tolerate quite large amounts (Herbert and Mann, 1958). There is, however, as yet very little information available, this being another aspect of pollution which would repay further study.

Pollution of lakes

More important than the common biologically inert salts, such as sodium and calcium chlorides and magnesium and calcium sulphates, are the nutrient salts needed for plant growth. The principal ions involved are potassium, nitrate and phosphate, of which the last two are the most important as they are often in short supply in natural waters and so control the amount of plant growth which is possible. We have seen that the release of these ions from organic matter has profound effects on plant growth in rivers, causing massive developments of small algae, *Cladophora* and *Stigeoclonium*. This effect is of far greater consequence in lakes.

A lake into which an organic effluent flows, either directly or via a river or stream, receives an increased amount of nutrient salts, and this is so even if the organic matter is fully mineralised by the time it reaches the lake. In practice this is the only important aspect of lake pollution because, even where imperfectly treated effluents are allowed to flow into lakes, the dilution is normally so great that other results of pollution, such as the effects of poisons or the growth of sewage fungus, are usually strictly localised (Thomas, 1955b). It was explained in Chapter II that nutrient salts tend to accumulate in lakes, which become more eutrophic as a result. Clearly this process is speeded up by the addition of extra nutrients, and there is evidence that this is happening to lakes in many parts of the world (Hasler, 1947).

The results of increased eutrophy are almost all undesirable. We have seen that the more favoured food-fishes are replaced by a succession of coarse fishes, and that in Europe this means that trout, char and lake herrings are succeeded by perch and pike, and these in turn by bream and roach. The production of planktonic algae increases, as may the growths of weeds. The water becomes turbid because of the massive growths of algae, it loses its sparkle and colour, and some of the blue-green algae which flourish under these conditions are poisonous to fishes and stock (Prescott, 1948; Ingram and Prescott, 1954).

Cattle, sheep, pigs and many kinds of poultry, including ducks, have been known to die soon after drinking water that contained heavy algal growths (Fitch *et al.*, 1934), and there are many reports of deaths of fishes caused by the decay of algal blooms (Mackenthun *et al.*, 1945). These are not caused entirely by oxygen depletion, but by the release of toxic substances, especially by blue-green algae which are particularly characteristic of eutrophic water (Prescott, 1939). Other consequences of increased eutrophy are the decline in suitability of the water for industrial use and water supply, and a general loss of attractiveness from the recreational point of view (Lackey, 1949). Bathing becomes unpleasant because of the development of weeds and scummy blooms of blue-green algae such as *Anabaena* (*Fig.* 2), and some of the algae which develop under these conditions may render the water blood-red. *Oscillatoria rubescens* is one of these species, and doubtless one of them was responsible for the first plague of Egypt. In addition many of these algae produce unpleasant smells and tastes (Whipple *et al.*, 1947., Prescott, 1948), and storms deposit them, together with weeds, on the shores, where they decay in unsightly and malodorous masses.

Lackey (1945) has related the production of algal blooms in Wisconsin lakes to the concentrations of nutrients in the water and hence to the amount of sewage entering the lakes, and he has shown (Lackey and Sawyer, 1945) that some of them receive annual dosages of saline nitrogen ranging from 73–422 lb./acre, which is many times the amount normally applied to farm land. It appears that about half the nitrogen is built up into organic matter in these lakes and that there is also adequate phosphate for this enormous amount of plant growth, the wet weight of which would be at least 100 times as much as the amount of nitrogen used. Even if nutrient salts are added while still bound up in organic matter they become rapidly available for algal growth (Flaigg and Reid, 1954; Ohle, 1955), so it makes little difference if they are added as purified or unpurified effluents, although of course ordinary biological treatment does remove some saline nitrogen and phosphate by sedimentation. Ohle (1955) states that raw sewage sometimes contains as much as 15 mg./l. of phosphate phosphorus, but treated effluents contain usually only 2–4 mg./l. although as much as 6–8 mg./l. may remain.

In a recent study of a large lake near Copenhagen (Berg *et al.*, 1958) it has been calculated that, because of pollution, about 24 tons of saline nitrogen and 4 tons of saline phosphorus enter the water each year, and that this represents about 12 per cent. of the total amount used by the plankton. Moreover very little of this nitrogen and phosphorus leaves the lake via the outflow, the calculated amount being about 3½ tons of nitrogen and 200 lb. of phosphorus. This emphasises the fact that lakes are very efficient traps of fertility, and that even slight pollution is likely to cause a rapid increase in the rate of ageing.

Unfortunately the change seems to be irreversible—once a lake has become eutrophic it remains so, at any rate for a very long time, even if the source of extra nutrients is cut off (Hasler, 1947). Another unfortunate feature is that the onset of extreme eutrophy appears to be a rather sudden feature in lake development, which takes only a few years to become manifest. Its appearance therefore tends to take the general public by surprise. Hasler records several examples of the sudden appearance of blooms in lakes in which they had never before been observed. The reasons for this are complex but as they are fundamental to an understanding of lake biology they will be reviewed very briefly here; for fuller information the reader is referred to Welch (1935), Macan and Worthington (1951) and Hasler (1947).

Water, unlike most liquids, is denser at a temperature of 4° C. than it is at higher or lower temperatures. As a result this is usually the minimum temperature reached by the main body of water during the winter: any water which becomes colder floats at the surface until it again becomes warmer or freezes. In the spring the upper layers become warmer, but also lighter, and as the summer advances the wind is less able to mix this warmer water with the colder layers below, and a distinct thermal discontinuity is formed. This so-called thermocline lies usually at depths of the order of 50 ft., but the exact depth varies with the storminess of the spring and the shape and situation of the lake basin. The upper water, or epilimnion, remains in full contact with the air and is circulated freely by the wind, but the lower cold layer, or hypolimnion, is cut off from all sources of oxygen. It has no contact with the air and is too dark, because of the absorption of light in the epilimnion, for plant growth. It therefore has only the reserve of oxygen which it acquired during the period of full circulation in the spring.

In an oligotrophic lake there is little oxygen demand in the hypolimnion because of the general paucity of life and the absence of much organic matter sinking from above. The store of oxygen is therefore sufficient to last until the autumn, when complete mixing again occurs because of the cooling of the epilimnion. In a eutrophic lake on the other hand there is a large oxygen demand in the hypolimnion because of the constant rain of dead and dying plankton, and all the oxygen is used up during the summer at least near the bottom. This of course has marked effects on the benthic fauna, which do not concern us here, but it also affects the release of nutrients from the dead organisms. Under aerobic conditions these salts tend to remain in the mud, and relatively small amounts of them find their way back into the water; under anaerobic conditions, however, they are released very readily into solution and hence, ultimately, back into the biological cycle.

Therefore, as a lake reaches that state of productivity which results in total de-oxygenation at the bottom of the hypolimnion it becomes considerably more productive, and may begin to produce plankton blooms quite suddenly. It is at this stage that the general public becomes aware that the lake has changed, and within a very few years there may be marked losses of amenity.

To the limnologist, of course, the change does not seem so sudden because the drastic effects are preceded by more subtle chemical and biological alterations. For instance the total salt content of the water rises slowly—this can be shown by measurements of its electrical conductivity; the oxygen content of the hypolimnion in the late summer becomes lower and the turbidity of the water increases as the amount of plankton increases. It has been suggested that increased turbidity in the Great Lakes of North America may be a factor in the decline of fish populations, but Van Oosten (1945) has denied this and maintains that the decline is entirely due to over-fishing. It is almost certain, however, that changes are occurring in these lakes which must affect both the turbidity, because of increased production of plankton, and ultimately the fishing. In lakes which are known to be changing it is found that the dominant species of algae succeed one another in a fairly definite way.

Desmids, like *Cosmarium* and *Staurastrum*, disappear and are replaced by the diatoms *Asterionella* and *Synedra*, and these in turn are replaced by blue-green algae such as *Anabaena* and *Microcystis* (*Fig.* 2). Many of these changes have been recorded as having occurred in the past thirty-five years in Fure Lake near Copenhagen (Berg *et al.*, 1958), where it has also been found that the higher plants have become confined to shallower water, almost certainly because of increased turbidity.

In Switzerland also several lakes have been under continuous observation for a long period, and this applies particularly to Lake Zürich (Thomas, 1956/7). This extensive lake receives large amounts of nutrient salts from sewage and industry, and many changes have been observed in it during the last half-century. The colour of the water has altered from blue to yellowish or brownish green, because of the increase in plankton, and since 1896 plankton blooms have occurred. These have been caused by species which previously were absent or rare. *Tabellaria fenestrata* was the first to appear, followed by the red-coloured *Oscillatoria rubescens* in 1897. New diatoms appeared in the plankton for the first time in 1905 and 1907, new green algae in 1923 and 1945, *Anabaena planctonica* in 1947, *Lyngbya limnetica* in 1951 and *Rhizoselenia eriensis* in 1955/6. At the same time growths of littoral algae have increased to such an extent as to spoil bathing beaches. The oxygen content of the hypolimnion has fallen steadily, and this has caused trouble with water supplies which are drawn from the deeper parts of the lake. These are some of the warning signs, and there are others, many of which are given by Hasler (1947).

Even in these early stages of the process of eutrophy the water becomes less useful for many purposes, particularly water supply. Many of the planktonic algae produce tastes and smells, especially if they are killed by bruising or chlorination, and some species, most notably *Synura*, cause unpleasant tastes even when they are fairly sparse. For a fuller account of this aspect of the subject the reader is referred to Whipple *et al.*, (1947), Pearsall *et al.*, (1946) and Prescott (1948).

It is clear therefore that any increase in the rate of eutrophy, even though this involves only the acceleration of a natural and inevitable process is, from a human point of view, thoroughly undesirable. It is also obvious that pollution of lake water by sewage even if it is competely mineralised, must have just this effect.

In conclusion is should be noted that the effects of rapid eutrophy may occur also in narrow arms of the sea, and the results may be just as spectacular and undesirable as they are in fresh water. Braarud (1955) reports that excessive sewage pollution of Oslo Fjord has caused total de-oxygenation in the deeper water and the production of sulphides, and this has destroyed commercial trawling in the area. In the surface waters algal blooms now occur, and one species, *Coccolithus huxleyi*, is often so abundant that the water becomes milky and unattractive. More serious, however, are occasional blooms of Dinoflagellatae. Many of these are poisonous and on being eaten by shell-fish make these poisonous also; Braarud records that people have been killed by eating shell-fish from the fjord.

Chapter XII

OTHER HUMAN INFLUENCES ON NATURAL WATERS

APART from the alterations caused by effluents many of man's activities have profound effects on the living communities of inland waters, and although this book is concerned mainly with pollution in its restricted sense it is not out of place to consider some of these other influences.

It was pointed out in Chapter I that the clearing of forests increases the rate of run-off of surface water, and with it the liability to sudden spates. From the discussion in Chapter IV it will be clear that this must result in a general reduction in the numbers of creatures in the rivers. Clearing of forests also affects still waters, because the destruction of the trees and the baring of the soil releases nutrient salts which are washed into lakes and increase their productivity. Pearsall and Pennington (1947) have shown that in the bottom deposits of Windermere there is a zone laid down between about 1100 and 250 B.C. which is very rich in organic matter. They concluded that this is the result of increased plankton production caused by enrichment of the water due to destruction of forests by the first large human population in the area. Later deposits contain less organic matter, and this may be due in part to exhaustion of the nutrients in the soil, but is also probably largely caused by increased soil erosion and the consequently greater amount of inorganic material in the deposit. Deposits laid down in the last century once again contain large amounts of organic matter, and this they attribute tentatively to the development of the tourist industry and the construction of sewage systems draining directly into the lake. It is also worth noting that, as was stated in Chapter I, even reafforestation may produce biological effects on aquatic habitats when certain species of trees are planted.

Of more obvious importance are alterations made directly to the water bodies themselves. The effects of large dams are so manifest that they do not merit discussion, but small dams such as those made for mills or to produce sufficient depths for water intakes have less obvious consequences, especially in polluted rivers as we have already seen. They constitute barriers to migratory fishes, but this fact is now so well known that most are provided with fish-passes where these are necessary.

It is not, however, generally realised that dams and weirs form barriers to some invertebrates, which are unable to move upstream through fish-passes. Clearly only those species which have no aerial stages are involved and creatures, such as snails, which can crawl up wet surfaces are not seriously impeded. Also, of course, these barriers are important only where for some reason the species concerned are absent above the obstruction. This may result from past pollution, which has exterminated sensitive animals, or when there is a newcomer to the fauna.

Examples of each of these types are known. For instance *Gammarus pulex* was eliminated by pollution from the river Lee, from its source for at least 10 miles. This stretch of the river serves several water mills, and when the pollution was reduced and conditions again became suitable for *Gammarus* no shrimps reappeared. This was almost certainly because they were prevented from doing so by the mill dams; when they were introduced near the source of the river they established themselves and bred satisfactorily. *Crangonyx pseudogracilis* is a North American shrimp which was introduced into this country some time during the first quarter of this century. It is now well established in Britain and has become a common inhabitant of canals, along which it is spreading into new areas. It moved into North Wales along the Shropshire Union Canal from which it reached the river Dee at the point of origin of the canal near Llangollen. From there it has spread at least 20 miles downstream, but it has been able to move only a few hundred yards upstream to the foot of a weir about 4 feet high. This presents no barrier to salmon which pass up the river in large numbers, but the shrimps have been unable to surmount it for at least ten years.

Another hazard to which river creatures are exposed is canalisation for drainage purposes. In recent years many rivers have been straightened and canalised in order to prevent flooding, and this seems usually to be done with complete disregard of the biological consequences. The river bed is attacked with bulldozers and mechanical shovels, the entire environment is altered and most of the animals and plants are removed. Unless the whole length of the river is so treated re-colonisation is of course possible, but usually the river bed is so altered that it can never return to its original state. In place of a series of pools and riffles the depth and current, and hence the substratum, are made uniform and, as we have already seen, decrease in variety of microhabitat leads to a general biological improverishment.

Also, of course, alterations of this kind eliminate the special conditions needed for spawning by game fishes, and if the areas involved are extensive there may be a marked decrease in their rate of reproduction. A particularly glaring example of this is a small south-country trout stream which originally meandered across a clay plain and had pools and gravel riffles and scattered beds of weeds and burr reed, the latter growing on silty shoals. This varied environment provided suitable breeding sites for fishes and a wide range of microhabitats for a rich and diverse fauna. Then almost half of the length of the stream was dug out with mechanical shovels in the interests of land-drainage. All the gravel, which must have taken many centuries to accumulate because stones are scarce in that area, was dumped on the bank and the stream was reduced to a straight clay dyke of uniform depth and current speed. Shortly before the digging a ten-minute collection of invertebrates yielded 254 specimens of at least 23 species in a wide range of taxonomic groups. A year later the only vegetation to be seen was a dark film of diatoms (*Navicula*) on the clay, and ten minutes' diligent search produced only 25 invertebrates of 9 species. All the worms, leeches, molluscs, bugs, beetles, caddis-worms and mites had disappeared, and only a few shrimps, stoneflies, mayflies and chironomid larvae remained. Undoubtedly this stream has been utterly destroyed from an amenity point of view; it is no longer attractive to look at, and the only fish species which could spawn in it is the three-spined stickleback, but even that is doubtful because of the swiftness of the current.

In this instance the River Board was in my opinion obsessed with its function as a drainage authority and completely oblivious to its other responsibilities. It would not have been impossible to have improved the drainage without destroying the stream bed. As it is this river will no longer support trout, nor for half its course any other fishes, and one can predict that the damage will be virtually permanent. In this respect it is worse than any kind of pollution.

Rather similar but less devastating instances of damage to streams are reported from Scandinavia, where water courses are cleared and straightened in order to float out logs from the forests. The actual process of clearing destroys some of the fauna, and the character of the streams is so altered that grayling tend to replace trout. Here again collaboration between biologists and engineers would undoubtedly reduce the damage. Müller (1958) has suggested that the actual course of the stream should not be altered, that as much variation in habitat as possible should be preserved and that erosion with its consequent silting should be avoided.

Drainage schemes may also affect lakes where these are used for the control of floods or to provide make-up water in the outflowing river during dry weather. The biological damage is confined to the shore, and results from the large and often rapid fluctuations in level which any such scheme must involve. Violent fluctuations of shore-line are also, of course, the inevitable lot of any lake which is used as a reservoir, and for this reason the shores of such lakes are usually bare of vegetation and often almost without animal inhabitants. This paucity of life results from the fact that aquatic plants and animals cannot withstand long exposure to dryness. The animals are, however, able to move, and in a normal lake they do travel up and down the shore as the water level fluctuates (Moon, 1935); but this movement is fairly slow and suffices only for small changes in level. When the range of water levels is greatly increased they inevitably become stranded and perish, and after a few violent fluctuations most of the littoral fauna is destroyed.

Unfortunately, although the deep-water fauna and the plankton are unaffected by this process, the shore is, in most lakes, the area richest in animal life, and it is very important as a forage area for many desirable fishes, amongst which are trout and perch. Loss of the littoral fauna inevitably, therefore, reduces the numbers or the growth-rate of these fishes, although this last point has not, I believe, actually been demonstrated. As an example of this kind of damage I may cite Llyn Tegid (Lake Bala) in North Wales, which in 1955 was provided with a barrage which enabled the water level to be controlled between heights ranging from its original highest level to 8 feet below its lowest level; a total range of about 14 feet. This was in order to control flooding in the valley of the outflowing river, the Dee, and to provide make-up water in dry weather, as the river is used for water supply. The shore fauna was originally rich and varied, but within a few months of the start of operation of the scheme it had been seriously depleted. All the sponges, molluscs, mayflies, stoneflies and caddis-worms had disappeared, together with most of the littoral vegetation; and other creatures, such as *Asellus*, *Gammarus* and *Sialis* which used to be very common, could be found only after prolonged search. The shore fauna had, in fact been reduced to a few species of worms and chironomid larvae, which presumably are able to escape desiccation by burrowing, and the little water boatman, *Micronecta poweri*, which is an active swimmer and so can keep pace with the changes in water level.

Moreover, during the first three years of operation of the scheme it seems that many of the eggs of the gwyniad, *Coregonus pennanti*, were destroyed. This fish, like so many of the lake herrings, is peculiar to that one lake, and used to support a local fishery. It feeds in the open water, but it spawns on the stony shores in shallow water during the winter and the eggs take several weeks to develop. In each of the three years much of the spawning occurred shortly before the water level was lowered, presumably to make room for spring floods, and eggs were stranded and killed. It will be clear that not many years of such treatment must inevitably exterminate a unique, interesting and at one time commercially useful fish. Here again a little co-operation between biologists and engineers could probably prevent a biological disaster without seriously disrupting the drainage scheme. Most of the damage done by drainage engineers seems to result from their ignorance of the biological results of their activities.

Agriculture affects inland waters in two ways. Cultivation leads to loss of soil and so increases silting, and fertilisers are leached out into streams and thence to lakes where they increase the rate of eutrophy. These points have already been discussed with reference to pollution and deforestation, but an important point about the leaching out of nutrients from the soil is that it occurs more readily when there is no vegetation (Jaag, 1955). Thus even if no fertilisers are imported on to agricultural land the very fact that it is bare of crops for some part of the year increases the supply of these salts to inland waters.

A recent human activity which has caused concern to those interested in aquatic life has been spraying or dusting with modern insecticides in order to control insect pests. This is often done from the air over wide areas of crops or forest, and it is inevitable that the insecticides ultimately reach rivers and streams. The chemicals most commonly used are D.D.T., B.H.C. and Aldrin, and they have caused deaths of fishes and food animals in Africa (Pielou, 1946) and North America (Young and Nicholson, 1951). In Switzerland it was found that dusting with B.H.C. caused even the tap water to be lethal to insects (Jaag, 1955). These insecticides are very toxic to fishes, and concentrations of B.H.C. and D.D.T. of the order of 0·04 mg./l. have been found to be lethal (Paul, 1952; Everhart and Hassler, 1945), but the toxicity seems to vary with the method of application. D.D.T., for instance, is less toxic when applied as a dust than as an emulsion, possibly because of the added toxicity of the carrier in the emulsion (Everhart and Hassler, 1945). Nevertheless this compound is so toxic that even the application of as little as ½ lb. to the acre of forest has killed young salmon (Kerswill and Elson, 1955). Moreover these compounds are more toxic to insects than they are to fishes; the yellow-fever mosquito, for instance, is killed by 0·005 mg./l. of D.D.T. (Hoffmann and Surber, 1945). Very low concentrations do not actually kill *Chironomus* larvae, but they so upset them that they are unable to build tubes, and this fact has been used as a measure of very small amounts of B.H.C. and D.D.T. For the former the critical point lies between 0·004 and 0·005 mg./l. and for the latter just below 0·02 mg./l. (Tomlinson and Muirden, 1948; Tomlinson *et al.*, 1949).

Although deaths of aquatic insects as the result of aerial spraying have been reported from several parts of the world, detailed investigations seem to have been made only in North America. Hoffman (Hoffmann and Surber, 1945; Hoffmann and Drooz, 1953) investigated streams in Virginia and Pennsylvania before and after the spraying of their watersheds with D.D.T. He found that many aquatic insects were exterminated, while others were severely reduced in numbers. The groups most affected were mayflies, stoneflies, caddis-worms, beetles and buffalo gnats, but even within these groups some species were less affected than others. This is of course a general phenomenon common to all forms of toxic pollution. Snails, fishes, dragonflies and chironomids were unaffected at the concentrations applied. It will be appreciated, however, that the biological communities of these streams were seriously impaired and, although in each case there was only a single application of poison and fairly rapid recovery could be expected, such treatment over a very wide area could have long-lasting effects. He also noted that aquatic mosses accumulated the poison, and so presumably remained poisonous for some considerable time.

In this country large-scale aerial spraying is not a normal activity, but it may become so as the areas of forest increase, and orchards and some field crops are regularly sprayed from the ground. There have been no serious complaints of damage to streams here, but it is worth recording that I observed that certain species of insects were always absent from one reach of a tributary of the Thames, although present further upstream. There seemed to be no ecological reason to account for this, but it is significant that a large chemical firm manufacturing insecticides maintains an experimental station and store on the river bank just where the affected reach begins.

Another disturbing development is the recommendation, which has been made repeatedly, that sewage works should use D.D.T. or B.H.C. to control the flies which breed on biological filters (Tomlinson, 1945; Tomlinson and Muirden, 1948; Tomlinson *et al.*, 1949; Wilson, 1949). These flies are a nuisance, as they often appear in clouds and enter houses, and it is possible that they could carry disease with them. Application of the insecticides has been found to be effective, but these substances are not destroyed in the filters although they are absorbed and released slowly. In one series of trials (Tomlinson *et al.*, 1949) the effluent sometimes contained over 0·1 mg./l. of D.D.T., and even 11 days after treatment 0·008 mg./l. was present. These amounts were not toxic to trout, because most were below the toxicity threshold and those that were not did not persist for long enough to kill fishes. They would, however, undoubtedly be sufficient to kill many normal aquatic insects. This practice does therefore represent a very definite danger, and should not be indulged in without very careful consideration and experiment.

It is also the practice, particularly in reservoirs but also in other lakes, to control algal blooms by the application of algicides, usually copper sulphate, to the water. Details of this process are given by Whipple *et al.* (1947), but it has been found that the doses required are less than those suggested there (Domogalla, 1941; Moyle, 1949), and nowadays fractions of a part per million are applied with power-spraying equipment. Such amounts control algae very effectively, and it is thought that they do not affect other plants or animals (Moyle, 1949). Indeed it has even been claimed that fishes in a Wisconsin lake benefited by the treatment when, in 1936, they were suffering from an epidemic of fungus disease (*Saprolegnia*) which ceased when the annual spraying programme began (Domogalla, 1941). In that lake treatment was found to be so unexceptionable that heavier doses were used, under carefully controlled conditions, to eliminate weeds and sessile algae from bathing beaches. Other investigators are not, however, so sanguine. For instance Moyle (1949) points out that although long-term treatment with copper sulphate, even at the relatively high concentrations needed to kill snails, has apparently caused no ill effects to fishing success, deaths of fishes do occur occasionally. Some of these may be due to de-oxygenation caused by the sudden decay of large numbers of algae, a phenomenon which sometimes occurs quite naturally (Whipple *et al.*, 1947), but, as with all heavy metals, the toxicity of copper is greater in softer water. It is likely therefore that dosages which are safe in some lakes would be dangerous in others. Undoubtedly, also, the copper, which becomes chemically united with the algae (Prescott, 1948), accumulates in the mud, and this may affect benthic animals (Lackey, 1949).

More recently sodium arsenite has been suggested for the control of weeds (Mackenthun, 1950), with doses up to 10 mg./l. This is claimed not to affect fishes, although some invertebrates, including shrimps, mayflies and Chironomidae, may be killed. Control of weeds, however, involves only the local areas where they are actually growing. These methods therefore need to be applied with care and with due consideration of what is being done. Obviously it is a question of weighing up one aspect against another. In a waterworks reservoir control of algae is of paramount importance, but in a lake which is used for amenity purposes chemical control of weeds and algae may benefit yachtsmen and bathers at the expense of fishermen. Here again biological investigation can at least clarify the issues between which choice has to be made. Fortunately in England, where summer temperatures are relatively low, excessive growths of weeds or blooms of objectionable blue-green algae are rare (Whipple *et al.*, 1947), so this human aspect of the matter is of less urgency than it is for instance in parts of Europe or the United States.

A further point which has to be considered in the chemical control of algae is that some species are more resistant than others to the poisons. Resistant species may survive and produce further, uncontrollable, blooms shortly after the treatment (Whipple *et al.*, 1947; Pearsall *et al.*, 1946) because of the sudden release of nutrients from the algae which have been killed. For this reason it is general waterworks practice to watch the algal population very closely throughout the summer and try to forestall blooms by early treatment. This is not too difficult in reservoirs, but would be an awkward and expensive practice to apply to other lakes which are used primarily for recreational purposes. One can imagine also that it would be difficult to convince fishermen that it was either necessary or desirable.

A practice which is probably quite as risky to the normal life of lakes and ponds as the chemical control of algae and weeds is the selective poisoning of fish with derris. This has been widely used in America for the control of fish populations and the elimination of undesirable species before restocking programmes (Leonard, 1939; Greenbank, 1941). This substance was originally used as a fish poison by the natives of Malaya, but it also has powerful insecticidal properties, and before the discovery of the modern synthetic poisons it was widely used in pest-control. Very little is known about its effect on aquatic invertebrates, but it seems very likely that it affects them adversely at the concentrations needed to kill fishes. The method has been used to some extent in Britain in order to clear lakes of coarse fishes before stocking them with trout. This is probably an unwise thing to do unless it is planned to leave the insect population time to recover before the introduction of new fishes, and in any event there is danger of damage to the outflowing stream. Netting or electric fishing are undoubtedly safer methods of eliminating undesired fishes, although they are more difficult to use in lakes and may be less effective than derris.

Lastly in this brief survey of some of the human activities which may affect the life in natural waters we may consider the recently developed method of preventing excess evaporation from lakes, ponds and reservoirs by covering them with films of fatty alcohols. These have been used with success in Australia, Africa and America, and although their chief importance will continue to be in arid or hot lands it seems not unlikely that they may also come to be widely used in Western Europe, where the increasing demands on natural water supplies are becoming more and more difficult to satisfy. Cetyl and stearyl alcohols, often mixed together, are applied to the water surface, on which they form monomolecular films and so reduce the amount of evaporation. Much water is thus saved, and although the alcohols have to be constantly renewed the amounts involved are so small that the practice has proved well worth while. To date, however, we know little about the biological effects of the film. It does not apparently greatly impede the passage of oxygen, nor does it interfere with plants or fishes, but there is some evidence that it affects insects which emerge at the surface. It has for example been found that mayflies are unable to emerge satisfactorily through the film (United States Department of the Interior, 1957), and it seems likely that the same may apply to Chironomidae, whose larvae are important inhabitants of pond and lake bottoms. Clearly, therefore, further study is required of the biological effects of these alcohols before it is possible to predict how they will alter the life in lakes on which they are used, and once again these investigations will require the co-operation of biologists and other scientists.

The conclusions of this chapter are therefore that any interference with the normal condition of a lake or a stream is almost certain to have some adverse biological effect, even if, from an engineering point of view, the interference results in considerable improvement. At present it would seem that this is little realised and that often much unnecessary damage is done to river and lake communities simply because of ignorance. It is of course manifest that sometimes engineering or water-supply projects have overriding importance; and even if they have not, the question of balancing one interest against another must often arise. But, regrettably, even the possibility of biological consequences is often ignored. It cannot be emphasised too strongly that when it is proposed to alter an aquatic environment the project should be considered from the biological as well as the engineering viewpoint. Only then can the full implications of the proposed alteration be assessed properly, and a reasonable decision be taken. Obviously this will vary with the circumstances and the relative importance of the various consequences involved, but, at present, unnecessary and sometimes costly mistakes are often made because the importance of biological study is unknown to many administrators. Often, as for instance in drainage operations, it would be possible to work out compromises which would satisfy both engineering and biological interests.

Chapter XIII

THE BIOLOGICAL ASSESSMENT OF POLLUTION

POLLUTION is essentially a biological phenomenon in that its primary effect is on living things. The same can also be said of the various other types of alteration to natural waters discussed in the last chapter. Nevertheless, despite its essentially biological character, it is the general practice to assess pollution in purely chemical terms, by measuring such things as dissolved oxygen, B.O.D., suspended solids and ammonia, and often no particular account is taken of the details of biological effects. If fishes are killed or large quantities of sewage fungus are produced these facts may be noted, but they usually occupy a subsidiary position, and because engineering and similar activities make no permanent alteration to the chemistry of the water it would seem that they are generally considered to be unexceptionable.

One has only to study the standard books on water to discover how little attention is paid to biology in the study of pollution. Amongst these I may cite the British book by Suckling (1944), the American one by Phelps (1944) and the German one by Olszewski and Spitta (1931), each of which contains a great deal about the chemistry of water, together with details of chemical methods, but either very little or very imprecise information on biological aspects. Bacteriological methods are, however, described in detail in two of these books (Olszewski and Spitta, 1931; Suckling, 1944), but mainly with reference to the measurement of total numbers of bacteria by plate-counts, and the detection of *E. coli* and some species which are pathogenic to man. These tests, although strictly speaking biological, are closely akin to chemical determinations.

This would seem to be an odd state of affairs when one considers that, apart from public health, man's primary interest in pollution, or in any other alteration to a natural water, is in its effects on general amenities and fishing or on the subsequent use of the water for other purposes. These are primarily biological in character, and even if the water is to be used for cooling purposes only its capacity to grow sewage fungus is of much greater importance than its B.O.D.

To some extent the reasons why chemistry has consistently taken precedence over biology in this particular field are historical. The early work on pollution, particularly that of the Royal Commission, was primarily chemical because, in the first few years of this century when the problems involved were becoming clearer, chemistry was already an established science whereas there had been little systematic study of freshwater biology. Chemical methods therefore became firmly established and, particularly in this country but less so on the Continent and in America, biological investigation was thought to have little to offer. The idea seems, also, to have arisen that biological phenomena are difficult to investigate and that their interpretation requires a highly skilled specialist (Klein, 1957a).

The implications of this would appear to be that chemical data are much easier to assess, and that the chemists involved in this type of work need not be particularly skilled. The information given in Chapters V and VI will, I hope, convince the reader that this is very far from being true. Interpretation of chemical data is every bit as difficult as is that of biological data, especially when it is appreciated that chemical analysis is at best an indirect measure of the basic biological properties of the water. To know the oxygen content, B.O.D., and any other measurable chemical or physical properties of an effluent, can only allow one to guess at the effect it is likely to have on the living things in a river; only direct biological study can actually determine what these will be.

Biological study has moreover several advantages over chemical analysis. It takes less time, because a single series of samples reveals the state of the animal and plant communities, which themselves represent the results of a summation of the prevailing conditions. The chemist has to make a long series of observations in order to obtain an average value, and even then his samples may all happen to be collected at times other than those of extreme conditions, which are often the most important. Biological investigation also shows up the effects of intermittent pollution or the result of a single discharge of a poisonous substance, which the chemist may miss altogether, and it enables the investigator to pin-point the source of the trouble. To do this the biologist has merely to move upstream until he reaches the place where the biological effects begin. This is possible for the chemist only if the pollution is continuous, and even so must involve him in the analysis of a large number of samples.

The advantages of biological investigation therefore lie in the fact that the animals and plants provide a more or less static record of the prevailing conditions and that they are not affected by a temporary amelioration, nor usually by a transient deterioration, of the effluent. But biological investigation does not have all the advantages. It is to some extent a true indictment to say that its results are not always easy to interpret, although I hope that this book will show that the difficulty is often exaggerated. It is, however, perhaps worth stressing that it takes a trained biologist to interpret biological data, just as it takes a trained chemist to interpret chemical analyses. No book or short course of study can turn an untrained man into a freshwater biologist any more than it can turn him into a water chemist; both must acquire their essential background knowledge by training and experience.

Biological study can reveal only the general type of the pollution; it does not indicate the exact substances involved. It can distinguish between organic and poisonous pollution, but it does not identify, except in certain cases, the particular poison which is causing damage. That remains the province of the chemist, although the biologist can often indicate to him the sort of analysis which is worth making. Biological analysis also has the great disadvantage that it does not deal directly with concentrations. The sewage-works manager and the industrialist need figures to which they can work; they must have some sort of measure by which they can test the quality of their effluents and so check the efficiency of their purification plants. This is most readily supplied by chemical analysis and they therefore turn to the chemist for information about permissible concentrations. Herein lies the major dilemma of the scientists concerned with pollution. We simply do not know what concentrations are permissible, and even though figures are often quoted they are in reality only the result of informed guesswork.

It will be clear from Chapter VII that there is very little information which can be used to determine how much of any poisonous substance can be allowed to enter a stream without affecting the living community. We are a little better informed about organic matter because under certain circumstances it is possible to calculate its effect on the oxygen régime of the river (Phelps, 1944), but, as was pointed out in Chapter VI, these calculations are full of uncertainties, and we know that de-oxygenation is not the only effect of organic pollution. We are also ill-informed about the minimum oxygen concentrations which are tolerated under natural conditions. Such figures as the 20/30 Royal Commission standard (B.O.D./suspended solids in mg./l.) are therefore fundamentally arbitrary, and are valueless without due consideration of dilution and other local factors. It is in fact fairly clear that this widely used standard is not sufficiently stringent, even with the eight times dilution for which it was postulated, to preclude biological alteration in a clean river. The confidence given by the numerical exactitude of chemical data is therefore spurious, and it is likely to remain so for a long while. But the more that chemists and biologists collaborate in this important aspect of pollution the easier it will be to work out what concentrations really are reasonable. Each effluent will, however, always remain a special case, because each river, and indeed each reach of each river, is different and so requires individual consideration.

Even now, when freshwater biology is an established science and capable of making a great contribution to the assessment of pollution, one still finds biologists making appeals that more use should be made of their methods (Liebmann, 1942; Surber, 1953; Patrick, 1953). Personally I am in full agreement with these appeals, but I also feel that the neglect of their viewpoint has been in some measure the fault of the biologists themselves. Their subject is a difficult one, and it is much less easy to explain to the layman than is chemistry. In their attempts to do this and make their methods generally available they have sometimes tended to over-simplify, while at the same time forgetting that the layman is very soon discouraged by long lists of scientific names. This was a particularly unfortunate thing to do during the first few decades of this century when many members of the general public, and indeed many scientists, were disinclined to take biology seriously. To them the biologist seemed to be a slightly ridiculous figure with a butterfly net, and when they found that his analytical methods were incomprehensible to them they, somewhat naturally, turned again to the familiar and apparently more precise methods of the chemist.

This state of affairs was, however, more marked in the English-speaking countries than it was elsewhere, and serious study of the biological aspects of pollution was begun in Germany at an early date (Liebmann, 1951). The results were first codified by Kolkwitz and Marsson (1908, 1909), who developed their now well-known 'Saprobiensystem' for the assessment of organic pollution. But perhaps well-known is too strong an epithet to use, because although it is widely used on the Continent it is rarely even mentioned by American or English authors. Kolkwitz and Marsson postulated that when a river received a heavy load of organic matter the normal processes of self-purification would result in a series of zones of decreasingly severe conditions succeeding one another downstream, and each containing characteristic animals and plants. These zones they defined as follows:

Polysaprobic: the zone of gross pollution with organic matter of high molecular weight, very little or no dissolved oxygen and the formation of sulphides. Here bacteria are abundant, as are other organisms, but there are few species of animals and these all live on decaying organic matter or feed on bacteria.

Mesosaprobic: with simpler organic molecules and steadily increasing oxygen content. This zone is divided into an upper (α) zone, still with many bacteria and often fungi, and with more types of animals but few algae, and a lower (β) zone where mineralisation has proceeded further and conditions are suitable for many algae, and tolerant animals and some rooted plants may occur.

Oligosaprobic: the zone of recovery where mineralisation is complete, the oxygen content is back to normal and a wide range of plants and animals occur.

They list a large number of species of plants and animals which are said to be characteristic of each zone, and suggest that identification of those which are present in a river or lake will enable its pollutional status to be assessed. If the load of pollution is not very heavy the upper zone will be α- or β-mesosaprobic or even oligosaprobic.

This system, although it clearly has some validity, has been subjected to much criticism, but it has also been extended and developed by many later workers. The criticisms were mainly directed against the placing of certain organisms in the particular zones, and against the identification of some of the animals. Kolkwitz was a botanist and Marsson a microbiologist, so it was not surprising that their knowledge of the larger animals was deficient. It is, however, unfortunate that some of these mistakes were repeated in 1950 when Kolkwitz published a revised account of the system, to which ecological notes on many of the species were added. Some workers have criticised the lists as containing many organisms which may be found in a wide range of conditions. For instance Fjerdingstad (1954) has shown that the flagellate *Bodo minimus*, which Kolkwitz classifies as α-mesosaprobic, is really tolerant of a wide range of conditions, but as it is unaffected by hydrogen sulphide it tends to be free of competitors in polluted water and so to become common. As, however, it may occur in other places its mere presence is of little use in the indication of pollution.

Other workers have stated that simple lists of the species present are insufficient for the assessment of the condition of the water. They conclude that the mere presence of certain species gives very little indication of the condition of the habitat, and that what matters is the whole community. They also stress the fact that changes in numbers of individuals, in the species which are dominant, and in the basic way of life of the dominants, from for example saprophytic to photosynthetic, are of greater significance than mere changes in the list of species present (Hentschel, 1925; Nowak, 1940; Liebmann, 1942; Šrámek-Hušek, 1958; Gaufin and Tarzwell, 1956).

With these various points in mind Liebmann (1951) has drawn up a revised system, in which the organisms are more carefully selected and are described and figured. He has, however, retained the basic idea that the drawing up of lists of indicator species is sufficient for diagnosis, although he has notes on abundance and the particular conditions under which the various organisms thrive. His work therefore represents a considerable advance on that of Kolkwitz (1950) published only a year previously. It retains, however, the erratic treatment of the larger invertebrates, which was a flaw in the earlier system. Liebmann (1942, 1951) does not consider that they are as important as microscopic forms, on the grounds that the latter are more generally distributed, and that the smaller an organism the more readily it will react to chemical changes in its environment (because the ratio of its surface area to its volume is greater). He also points out that in the zones of severest pollution microscopic creatures are abundant whereas larger ones are generally scarce or absent. We shall return to this point later in this chapter.

Thomas (1944) and Šrámek-Hušek (1958) have taken the classification of the zones further than other Continental workers in that they have split the polysaprobic zone into three. Unfortunately they do not use the same terminology, and their zones are not quite the same. Thomas calls them α-, β- and γ-polysaprobic. The first of these contains mainly bacteria and a few ciliates, the second many ciliates, among which *Colpidium colpoda*, *C. campylum*, *Glaucoma scintillans* and *Paramecium* species are abundant (Šrámek-Hušek, 1958), and the third is the *Sphaerotilus* zone. In this zone Protozoa are abundant, but there are many more species and each is less common than are those in the β-polysaprobic zone, and fungi and Metazoa appear. Thomas gives a table showing the average chemical conditions he observed in the three sub-zones, but stresses that they may well apply only to the particular environment he was studying.

In America the concept of differing zones of pollution has been developed more or less independently of the European system, and accounts of the general ideas held there are given by several authors (Campbell, 1939; Brinley, 1942; Whipple *et al.*, 1947). In broad outline this concept is similar to that of Kolkwitz and Marsson, although the zones are less rigidly defined and do not quite correspond to those of the German system. They are in effect more of a descriptive nature and do not attempt to set up a strictly applicable system for biological analysis.

The implied claim of the Saprobiensystem, even in its modified form as proposed by Liebmann, to offer a hard-and-fast scheme for the classification of degrees of pollution is, I think, its major weakness. In the first instance the system is of course applicable only to organic pollution, and is quite useless for the assessment of the effects of poisons or other polluting matters. It also takes no account of the differing effects of various types of organic pollution. We have seen for example that some organic wastes stimulate the growth of sewage fungus without causing much de-oxygenation. The effect of heavy pollution of this kind is therefore very different from that of sewage. This is inevitable, because different creatures react differently to the various aspects of organic pollution. Some are adversely affected by low oxygen concentrations, and others are eliminated by the products of decay, amongst which are ammonia and sulphides which may not be produced at all by some organic effluents. Some species benefit by the rich food supply which is provided and yet others are affected by the change in the biological environment: as has already been stated the mere presence of a carpet of sewage fungus can have far-reaching effects. Inevitably therefore the zonation varies considerably and it is different for each type of effluent. Moreover the type of zonation produced is different in each type of river. If the water is shallow and turbulent even heavy pollution does not produce the de-oxygenation typical of the polysaprobic or, as the Americans call it, the septic zone. There is, however, a high content of complex organic compounds, and creatures which thrive on these substances are common. As early as 1918 it was found that the system was not applicable to the rapid rivers of Switzerland for this reason (Steinmann and Surbeck, 1918, 1922), and more recent workers have found that short turbulent reaches upset the smooth sequence of zones, allowing for example, the occurrence of such oxygen-sensitive creatures as stoneflies in zones which are otherwise classified as β-mesosaprobic (Nowak, 1940; Gaufin and Tarzwell, 1956).

Another weakness of the system is that the various zones differ according to whether they are primary or secondary. For example the α-mesosaprobic zone, when it occurs downstream of polysaprobic conditions, has by definition relatively simple organic compounds and large amounts of nutrient salts produced by the decay of the organic matter. It is therefore not a particularly good environment for bacteria or animals which feed on organic matter, but it is very suitable for the growth of tolerant algae like *Oscillatoria* and *Stigeoclonium*. If, on the other hand, the initial polluting load is lighter, the polysaprobic zone is absent and the α-mesosaprobic zone begins near the outfall, but its ecological status is quite different. Complex organic molecules are common and they encourage the growth of saprophytic plants and saprobic animals, but the amount of nutrient salts needed for the growth of green plants is relatively low and the typical pollution algae correspondingly scarcer. In other words the biological community is of a fundamentally different type.

For these reasons I consider that the Saprobiensystem, while providing a useful background of ideas and concepts, is rather a clumsy tool in the actual assessment of pollution in its many and varied aspects. To adhere to a rigid system of this kind is to bemuse oneself with the idea that complex ecological changes can be subjected to simple classifications. Nature is not as simple as this, and every example is different. The Saprobiensystem is applicable only to the particular conditions produced by heavy sewage pollution in a slow and evenly flowing river. If the effluent is not sewage, or if the river is turbulent, it breaks down.

Many observers have stressed the fact that so-called 'indicator species' must be used with extreme caution, as they often occur also in quite unpolluted water. A moment's reflection will, indeed, make it clear that this must be so as no species is *adapted* to living in polluted conditions, which are the products of civilisation and so very much younger than is any species. Absences of species may give just as much information about the condition of the water as the presence of others (Ellis, 1937), but in assessing the significance of such facts one must set them against a general knowledge of freshwater biology. It is important for the investigator to appreciate that, for example, most stonefly nymphs and caddis-worms are absent from the water during summer (Gaufin and Tarzwell, 1952). Richardson (1929), after his long and thorough study of the steady deterioration of the Illinois river, concluded that published lists of indicator species showed over-confidence, and that the occurrence of these species was so sporadic and variable that in the ultimate analysis 'the biological student is bound to be thrown back upon his own resources'. This is just another way of saying that it takes a trained biologist to make satisfactory use of biological data. He can explain his analysis to the layman, but he cannot produce a system of interpretation which the layman can use himself. This applies of course also to the chemist, but to the uninitiated this is less obvious, and the average layman believes, quite wrongly, that he could, if shown how, both carry out the chemical analyses and interpret the figures. The biologist, who occasionally tries to do this, has a greater respect for his colleague.

The Saprobiensystem, with its later modifications, is not the only formal system for the biological assessment of pollution which has been suggested. Patrick (1950, 1951) has described a complex method which involves the examination of a number of stations on a healthy part of the river and the counting of the number of species in each of seven arbitrarily chosen groups. The assumption is made that each group is a unit in that 'those species which are grouped together seem to behave in the same way under the effect of pollution'.

As each group contains a wide range of species and often several whole phyla, and some contain both plants and animals, this assumption seems to be quite unjustified, especially in view of the very varied reactions of quite closely allied species to poisons. However, having established how many species in each group occur in healthy reaches of the river, collections are made in the polluted reaches, and the numbers of organisms found in each of the seven groups are expressed as percentages of the numbers occurring in the healthy reaches. These percentages are plotted as histograms, and Patrick claims that the relative heights of the seven columns of the histograms give an indication of the degree of pollution. Quite apart from the very dubious assumption on which this system is based it seems to me that this is complexity for its own sake, and that biological findings can be expressed more simply and intelligibly. Kaplovsky and Harmic (1953), who have used this system, claim that it shows up slight pollution very well, but then so does the much simpler tabulation of numerical data (see Tables 8 and 11). They also state that a great deal of work is needed to establish the baseline in the healthy stations. I suggest that this time would be better spent in investigating the polluted region, and that, as has been stressed above, numbers of individuals are often as revealing as occurrence of species, especially where pollution is slight. A rather similar but simpler system has been suggested by Wurtz (1955), to which in some measure the same criticisms apply.

In conclusion I would stress that in my view it is a great mistake to try to evolve formal methods such as those quoted above. In nature little is simple and straightforward, and a rigid system can lead only to rigidity of thought and approach. Each river or stream and each effluent is different, so the pattern of pollution varies from place to place. But although the pattern varies the phenomenon is nonetheless detectable, as has I hope been made clear in previous chapters. If numerical data are collected and tabulated, or drawn as histograms, the effects of pollution are clearly shown even when it is very slight. There is neither need of, nor advantage in, a formal classification into zones, which in any event are not clearly defined, nor is anything to be gained by elaborate graphical methods. This has been particularly well illustrated by Butcher (1946b) who has shown how simple tables, showing the abundance of certain algae and invertebrates at a series of stations along the lengths of a number of British rivers, demonstrate the existence of various types of pollution.

There has been some dispute among biologists as to which type of organism is most indicative of pollution. This seems to me to be a very fruitless discussion. Obviously fishes are the least satisfactory because they are difficult to see or catch, and are less abundant than smaller creatures. They are also very mobile, so they often occur far away from their normal habitat. Bacteria, algae, rooted plants and invertebrates on the other hand are less mobile, more abundant and easier to collect, so they clearly offer greater possibilities; but whether one group or another of these is the most suitable subject for investigation depends on the circumstances. Ideally all should be studied, as all react to the various kinds of pollution.

Liebmann (1942, 1951) stresses the importance of micro-organisms in the assessment of organic pollution and points out that the inhabitants of severely polluted water are almost all microscopic; Šrámek-Hušek (1958) claims that the ciliates are the most useful animals among microscopic forms as they react sharply to organic matter. This is undoubtedly true, but micro-organisms suffer from considerable technical disadvantages. In the first place they are not easy to sample quantitatively unless they can be induced to grow on slides, as can algae. This difficulty has, however, been to some extent overcome by Šrámek-Hušek (1958), who has developed a precise sampling technique for ciliates which makes comparison between different stations possible. He stresses, nevertheless, that this technique has to be used with care as conditions such as current speed or oxygen content vary locally, so assessment must be based on many samples from each general locality. Secondly most micro-organisms must be examined alive or they are unidentifiable. This means in effect that laboratory facilities must be readily available, because even tolerant creatures will not live for long in grossly polluted water once it is allowed to stagnate in jars or collecting tubes (Liebmann, 1942). Thirdly it is very important, particularly with Protozoa, that identifications should be accurately made. For example, within the genera *Bodo*, *Vorticella* and *Paramecium* different species occur under quite different saprobic conditions (Liebmann, 1951). Unfortunately, specific identification of these minute creatures is far from easy and requires special training, which most freshwater biologists lack. Here then is the germ of the idea that biologists studying pollution need to be highly skilled specialists.

Lastly, while it is true that micro-organisms are very important in regions of gross organic pollution, they tell one, in practice, little more than can be determined by examination of such readily identifiable things as sewage fungus and algae such as *Stigeoclonium* or the diatoms, which have the great merit of remaining identifiable even after preservation in formalin. Furthermore we know very little of the reaction of micro-organisms to pollution which is not simply organic, nor do we know enough about them to make use of them in the assessment of very mild pollution.

Some investigators (Richardson, 1921, 1929; Gaufin and Tarzwell, 1952, 1956; Huet, 1949) have relied almost entirely on macroscopic invertebrates in their studies. These offer technical advantages in that they are easy to collect quantitatively and identify after preservation, so samples can be dealt with in a laboratory far away from the river. They are also particularly useful for the assessment of poisonous pollution and very mild pollution of all kinds (see Chapters VII and X). They have, however, the disadvantage that very few species, virtually only *Chironomus thummi* and the Tubificidae, can tolerate gross organic pollution. Thus, unless conditions of substratum are suitable for these animals or the flow is sufficiently sluggish to permit *Eristalis*, *Psychoda* or mosquito larvae to appear, they provide no satisfactory means of distinction between zones which might be classified as polysaprobic or α-mesosaprobic by the Saprobiensystem.

Butcher (1946b) has demonstrated that the algae growing on slides placed in the water give a clear indication of the severity of pollution by organic matter and to a lesser extent by metallic poisons. But here again our knowledge is as yet too limited for this method to be used for the study of very mild pollution.

In conclusion, therefore, it seems fair to say that all categories of living creatures are useful in the study of pollution, and that the investigator should use all the methods which are feasible for him under the prevailing conditions. For severe organic pollution, given that he has suitable laboratory facilities near at hand, ciliates and other micro-organisms are useful; for less severe cases or pollution of other types the larger invertebrates and the algae, both of which can be preserved and studied later at leisure, are useful. For mild pollution he can, in our present state of knowledge, use only the larger invertebrates. For all investigations numerical data and the study of the whole community are important: simple lists of species present, such as are advocated by the Kolkwitz and Marsson Saprobiensystem, are likely to be misleading.

Lastly we have to consider the relationship between the biological and the chemical assessment of pollution. Several European investigators have studied organically polluted waters both chemically and biologically, and have then compared the results obtained. The two methods have always been found to agree in outline, in the sense that both distinguish between clean water and severe pollution, but the detailed results have not fitted so well. Nowak (1940), studying a river in Czechoslovakia, found that local conditions in the river bed interfered with the smooth working of the Saprobiensystem, because shallow riffles allowed the appearance of sensitive creatures like stoneflies in reaches where chemical data indicated fairly severe pollution. This is of course merely an indication that the biologist must take note of local conditions in making his assessment. Thomas (1944) found that, even in his artificial experimental channels in Zürich, there were fairly wide chemical fluctuations in the zones in which particular biological communities developed. Liepolt (1953), in a very thorough study of a tributary of the Danube near Vienna, found that chemical conditions fluctuated widely during the 24-hour period and that they also varied from day to day. Biological data on the other hand gave a much clearer picture of the state of the river as long as the density of the organisms rather than their mere occurrence was considered. Similarly Huet *et al.* (1955), after a detailed study of two Belgian rivers, concluded that chemical data gave an indication of the biological state of the water only if they were interpreted with considerable care and a number of different chemical and physical factors were considered together.

These findings are not unexpected when one considers the complexity of the factors operating on the biotic community. The surprising thing is that the agreement is so good, and it is very much to the credit of the chemists that they have evolved methods which, although fundamentally very indirect, do enable them to assess fairly closely the biological condition of the water, which is, in the ultimate analysis, the thing that matters.

Even in the relatively simple environment of a static laboratory vessel the correlation between the two methods of analysis is not perfect. Nitardy (1942) kept sewage in covered and uncovered containers and studied the biological and chemical changes as it underwent self-purification. As in rivers he found only general agreement. The chemical data showed steady improvement, but no very clear relationship to the biological changes, which went through the usual sequences of bacteria including *Zoogloea* and *Beggiatoa*, to ciliates with *Sphaerotilus*, and thence to a more varied flora and fauna as the water cleared. And, of course, when the pollution is not a simple organic one the relationship between biological and standard chemical analysis breaks down altogether. Butcher (1955) has discussed the biology of the river Trent and its tributaries in relation to pollution, and has shown that chemical and biological data differ widely when several types of pollution are involved. He states that pollution should be defined 'not by any chemical standards, which can frequently be misleading, but by the actual changes brought about in the biological balance of the stream'. This I conclude is the outlook which should be adopted generally because, in the opening words of this chapter, pollution is essentially a biological phenomenon. This is not, however, intended to imply that chemical analysis is of little consequence. It measures an essential part of the environment, and when closely correlated with biological study it greatly enhances its value. When the chemist and the biologist both work on the assessment of pollution they can discover much more together than either can alone.

Chapter XIV

THE PROSPECTS FOR THE FUTURE

THE problem of effluent disposal is likely to remain with us as long as we live together in large industrial communities. As Klein (1957a) remarks, water pollution, 'like crime, disease and road accidents, will eventually be brought under control but will never be completely eliminated'. Effluents are an essential by-product of modern civilisation; they must be disposed of somehow, and in heavily industrialised countries they represent such a large proportion of the flow in the rivers that they can be returned only to inland waters. To pipe them away to the sea would result in dry river beds and serious water shortages but, as Phelps (1944) says, we must be intelligent in our use of streams, rivers and lakes for effluent disposal. Each is an asset to the community, and each has its particular uses, for fishing, boating, water supply, navigation or general amenity. Often these uses are mutually conflicting, and all of them may be adversely affected by effluent disposal. Set against this is the fact that natural waters are themselves extremely efficient in the actual treatment of waste material. By dilution, precipitation and oxidation they are able to purify surprisingly large volumes of effluent.

There is, however, a definite limit to the amount with which any particular water-body can deal without being seriously affected. The question of how each natural water should be used is therefore fundamentally one of administration, of a careful weighing up of the conflicting interests and, as far as rivers are concerned, of the condition of the water at various points. This, in Britain, is the duty of the River Boards, and there is little doubt that in time they will be able to discharge it effectively. Each will have to decide just what can and just what can not be allowed to enter its river at various points along its course, in order that a proper balance may be maintained between the various interests involved. Clearly the standards will vary from place to place and will have to be rigorous for some rivers, but it seems almost certain that other rivers will have to be given over almost entirely to effluent disposal. This does not mean that they need be allowed to become offensive or dangerously poisonous, but that they will be unsuitable for other purposes. It would seem for instance that some of the rivers of heavily industrialised parts of the country must come to be used in this way, but the loss of some rivers could be turned to advantage. They could for instance be used to relieve the load on others in the area, and their tributaries could be cleaned up at the expense of the main stream, so that aquatic amenities would be available even in the most densely populated areas. At present many industrial areas possess only polluted streams.

Planning for clean rivers must, however, go on at levels higher than those of the River Boards. New development schemes must take effluent disposal into account as well as other facts, such as suitability of site and water supply. There are hopeful signs that industry is becoming conscious of this point, largely because of a number of prosecutions, but it still often seems to be ignored by Government circles. For example at least two of the new towns have been sited on rivers which are manifestly too small to deal with the sewage which they will produce, no matter how well it is treated. As was stressed in Chapter V, very few effluents can, in our present state of knowledge, be purified to such an extent that they are devoid of any polluting effect. This applies with particular force to sewage which, even when completely mineralised and after the onset of nitrification, still retains some B.O.D. and inevitably contains large amounts of nutrient salts.

Improvements in effluent treatment are being worked out continuously, most notably by the Water Pollution Laboratory of the Department of Scientific Research (Southgate, 1951). Methods are now available for the improvement of most types of industrial waste, and in time they will doubtless come to be used universally. At present we are still suffering from the sprawling development of the Industrial Revolution and the aftermath of two world wars. The outlook is therefore hopeful; as time and money permit, old factories will acquire suitable treatment plant and new factories will be sited so that effluent disposal is possible without nuisance and damage.

Some problems will need to be tackled at their source. This applies particularly to radioactive elements, many of which are concentrated by living creatures, which could thus become dangerous even when grown in waters containing very little radioactive material. Fortunately, however, there is evidence that the need to consider pollution at the early stages of manufacture is now receiving consideration. For example Southgate (1957a) reports that investigations are being conducted on detergents with the aim of producing some substitute for the alkyl aryl sulphonates which are now in general use, but which cause foam on rivers because they are not destroyed by sewage-treatment plants. This work is being done solely because of the effects of these substances on rivers, and it demonstrates an increasing awareness of the fact that almost everything used domestically ultimately passes down the drain.

Careful thought must be given to engineering works which involve water courses. These, as we have seen, can do grave damage, but they can also alleviate pollution. When a polluted river is dammed a large volume of water replaces a small one, and the rate of passage through the valley is reduced. This allows more time for self-purification to occur and so reduces the severity of the pollution further downstream. In the Ruhr valley this method has been used as a substitute for biological treatment of sewage (Imhoff, 1931), and it could perhaps be widely used to supplement sewage treatment. Aeration also speeds up self-purification, and the construction of waterfalls at suitable points could doubtless often be arranged. Gameson (1957) has demonstrated that in many rivers more oxygen enters the water at weirs than in the reaches between them, and that the way in which the water falls has a great influence on the amount of oxygen taken up.

The most outstanding problem of the disposal of organic wastes, particularly sewage, is, however, the appalling waste involved. It seems to me that this requires much active thought in a world which suffers from the threat of food shortage. At present we are worried by the effect that sewage disposal has on rivers and lakes; we ought also to be considering its effect on the land on which we depend for our food. In round figures it requires the produce of one acre of land to feed one human being, and under primitive conditions there is a steady cycle of plant nutrients from the land to the human and back to the land. In civilised countries the plant nutrients are returned not to the land but to the water, and ultimately to the sea, where they are far less readily exploited. In this country we import about half the food we eat, so here we tip into our drains not only the fertility of about 25 million acres of Britain but that of about 25 million acres of other parts of the world. If we could recover some of that fertility we should not only preserve the rivers from degradation but we should be able to increase the yield of the land. It is of course an exaggeration to say that all this fertility is lost, because some is recovered in sewage sludge and settled solids, but the loss is nevertheless substantial. Every day many tons of potassium, nitrate and phosphate, which could be profitably used as fertilisers, flow to waste. The average human being passes 13 grams of saline nitrogen and 1·44 grams of phosphate phosphorus through his body daily. A high proportion of these, 88 per cent. and 50 per cent. respectively, is found in solution in raw sewage, and about three-quarters of the amount in solution passes through the sewage works (Thomas, 1956/7). On the basis of these figures one can calculate that the population of Great Britain wastes about 500 tons of saline nitrogen and 30 tons of phosphate phosphorus daily, and this takes no account of the losses in industrial effluents, many of which contain nutrient salts in large amounts. Nor does it include the many other sources of nutrient salts in domestic sewage: many detergents for instance are diluted with powders which contain phosphates (Ministry of Housing, 1956).

There is therefore an enormous daily waste of fertility and, as we have seen, this extra supply of plant nutrients causes trouble by stimulating the growth of algae. In rivers this is relatively unimportant, although large masses of *Cladophora* can be deleterious, but in lakes which are used for water supply serious difficulties may result. These have been discussed in Chapter XI, and because of their importance the exact factors controlling algal blooms have been much studied in both Europe and America. It has been shown that the addition of both saline nitrogen and phosphate to lake waters increases their production of algae, and that these two nutrients are the principal limiting factors (Lackey, 1949; Thomas, 1953), as there are normally adequate supplies of other materials needed for plant growth. In most lakes phosphate appears to be the ultimate limiting factor, as is shown by the fact that addition of nitrogen without extra phosphate does not increase algal growth, whereas addition of phosphate alone usually does so. Only in very eutrophic waters is shortage of nitrogen sometimes found to be limiting algal production. Springs, ground water and drainage water contain little phosphate (Thomas, 1955a), and it is not easily leached out of the soil, so even water draining from agricultural land contains only a few hundredths of a milligram of phosphorus per litre (Ohle, 1955). Saline nitrogen on the other hand is more plentiful and is usually present in milligrams per litre in natural waters (Thomas, 1955a).

The general consensus of opinion is therefore that the best way to control algal blooms would be to reduce the amount of phosphate entering lakes (Lackey, 1949; Thomas, 1953, 1955a, 1955b, 1956/7; Ohle, 1955). Fortunately it is possible to reduce the amount of phosphate in effluents by precipitation with alum (Lea *et al.*, 1954) or lime (Owen, 1953); doubtless other methods could be devised, but none is yet a general practice.

Imhoff (1955) has suggested that the nutrients in purified sewage could be removed by spraying it on to land, the fertility of which would then be increased by absorption into the soil; or by holding the sewage in shallow ponds where algae would be encouraged to grow and could be removed, together with their contained nutrient salts. A similar idea has been put forward by Pearsall (Anonymous, 1948) who suggests that water meadows or shallow lakes in which aquatic plants were grown could be a means of recovering fertility. Undoubtedly these are suggestions which need very serious consideration, and even if they prove to be impracticable it is to be hoped that they will stimulate thought about the possibility of preventing the enormous loss of land fertility which goes on at present.

In the sunnier parts of America algae are actually used in the purification of sewage, which is allowed to flow slowly through shallow ponds. The time required for purification is from 20 to 30 days, and sometimes some of the thick algal culture from the lower ponds is pumped back to the upper ones. This increases the amount of oxygen production where the B.O.D. is highest and thus speeds up the rate of purification. Ludvig and Oswald (1952) suggest that some such method might be used to produce a richly proteinaceous foodstuff. Algae can be precipitated and settled and could then perhaps be fed to animals. This promising idea has been discussed further by Isaac and Lodge (1958), who conclude that under some conditions it might prove practicable, although in Britain it seems probable that there is too little sunlight to allow of adequate algal growth. It seems likely also that there would be other complications. Steeman Nielsen (1955) studied the algal productivity of a Danish lake which was enormously enriched by nutrient salts from purified sewage. He found, as was to be expected, that the production was high, but it was not as high in terms of pounds per acre as well-manured barley fields in the same area. He concluded that the reason for this was shortage of carbon dioxide, which could not be absorbed from the atmosphere as fast as it was used up in photosynthesis. The pH of the water therefore rose at times to levels which were injurious to the algae.

Possibly this is a difficulty which would not arise with sewage effluents which were only partially purified, as they would continue to produce carbon dioxide from their own processes of decay, but it might be found necessary to introduce supplies of this gas by aeration or agitation. It is also possible that fertility might prove to be recoverable in the form not of algae but of sewage fungus. *Sphaerotilus* is very readily eaten by many invertebrates and some types of tropical aquarium fish. If it is suitable food for these animals it might prove to provide, when suitably treated, a valuable feeding stuff for pigs and poultry.

A more indirect method of recovering fertility is the culture of fish in ponds to which sewage is added. This stimulates algal and plant growth, and through them the production of invertebrates and fishes. This has been tried in Poland and Germany, where carp and tench have been reared successfully and yields of 300–500 lb./acre have been obtained (Kisskalt and Ilzhoffer, 1937; Macan *et al.*, 1942). In the European experiments it was necessary to dilute the settled sewage with clean water, and this clearly limits the places where such methods can be used, but in South Africa purified sewage has been used successfully without dilution. Hey (1955) reports that the effluent was found to reach oxygen stability after 12 days' storage in ponds, which were then stocked with plants. A large population of invertebrates then built up and fishes were added. In this way a yield of 1,000 lb. of fish per acre was obtained with no further feeding, and the fish produced were species of *Tilapia*. These, unlike European carp and tench, are first-class food fishes, and a yield of 1,000 lb./acre is of course very high indeed for a rich protein foodstuff. It would seem then that this also offers distinct possibilities of food production.

Unfortunately, both algal and fish growth depend to a large extent on climate, so such methods are likely to be much more productive in warm sunny lands than they are in Britain. Also, they require much land, and this is not available in the densely populated areas which produce the most sewage. These facts need not, however, be taken to indicate that the problem of recovering fertility is hopeless except in particularly favoured areas. In the promised atomic age it may prove possible to divert power to heat and illuminate ponds, and if the yield can be increased to a level that equals or even surpasses that of conventional agriculture—and there are indications that this may be possible (Isaac and Lodge, 1958)—shortage of land will cease to be of consequence. But this is perhaps an attempt to look too far into the future.

To return to present-day conditions, another item of waste in effluent disposal is heat. Every day millions of gallons of warm water are run into rivers, where they may do harm and certainly do no good. Recovery of low-grade heat is admittedly difficult, but is this not because the problem is always looked upon from an engineering standpoint? In Iceland large areas of glass-houses are heated by hot water from geysers, and they produce crops of tomatoes which could otherwise not be grown in the island. It would seem that a few acres of greenhouses would be a more profitable, and sensible, adjunct to a factory or power station than a number of cooling towers.

Power stations appear particularly suited to this sort of heat disposal as they are more active in winter when heat is most needed in greenhouses. Moreover they do not work to full capacity at night-time, and so have spare power available which could be used for illumination of the greenhouses and thus for the production of active plant growth in winter. Perhaps the reader considers this a wild suggestion, and it would certainly present some difficult problems, but it has been made before (Pirie, 1958), and if it could be made to work it would be a way of putting a waste product to good use. One might even be able to use the effluent of the local sewage works for hydroponic culture of the crops in the greenhouses. My main thesis, however, is merely that the problem of effluent disposal requires some radical re-thinking with special reference to the elimination of waste. I believe that biologists have a contribution to make in this field of thought, and that together with town and country planners, engineers and chemists they could work out advantageous methods of preventing pollution.

Returning once more to the present I wish to suggest that a task which is of immediate importance is an accurate scientific survey of the biological condition of all bodies of water which are liable to damage by pollution or engineering works. This need not be very elaborate, but it should be sufficiently detailed to put on record the present condition of the rivers and lakes.

In the past, when damage has been caused or suspected, it has usually been impossible to estimate what was the previous biological condition of the water, except by inference. If records existed the fact of increased or decreased damage could be easily and quickly estimated. Moreover such a survey would indicate places where very slight damage is already being caused (see Chapters VII and X) and so would show up danger spots, which could then be kept under close observation. Fundamentally this ought to be one of the duties of the River Boards and some of them have already performed it, at least in part. But many of them have no biologists on their staffs and still rely far too much on chemical data. Perhaps this work could be undertaken on a national scale by a small mobile team of biologists which could briefly survey each body of water in turn and then concentrate on areas which clearly require further study. As was pointed out in Chapter XIII such survey work can be done quite rapidly, and it would produce information of great value in the control of pollution and in the planning of sites suitable for further industrial or urban development.

Once pollution is brought under proper control it will be possible to consider the restoration of many of the rivers and lakes to their pre-industrial condition. Unfortunately, because of the irreversibility of eutrophy, it seems improbable that much can be done about lakes which have been prematurely aged, but it should be possible to restore rivers and streams to their original status. It is common knowledge that waters can be restocked with fishes, but haphazard restocking is not likely to achieve the best results. As we have seen, most normal river plants and invertebrates are affected by pollution and they, like fishes, may be completely eliminated. It is important, therefore, to make sure that they are given time to build up their populations before being exposed to predation by fishes. If this is not done the fishes will prevent or greatly delay the recovery of the normal river flora and fauna. If the original pollution involved only a part of the water course there will be a reserve of normal conditions from which recolonisation can occur, but if the whole river was affected it may be necessary to introduce such things as snails and shrimps whose powers of dispersal from one watershed to another are little, if at all, better than those of fishes. Once again I would emphasise that freshwater biology has a contribution to offer, and that rehabilitation should be supervised by scientists who would be fully aware of the consequences of any action taken.

Unfortunately, while it is true to say that once pollution has ceased it is possible to restore rivers to normal, this is only true in a general sense. Whatever is done the industrial revolution has done some irreparable damage. There is evidence that two and possibly three species of stoneflies which were confined to large lowland rivers have been exterminated in Britain (Hynes, 1958b). The same may be true of other groups of animals, because in almost all of them there are a few species which have not been recorded for a very long while, and which may belong to the original lowland river fauna. It would probably be quite possible to reintroduce these species from the Continent, where they still occur, as the Large Copper butterfly was reintroduced into Wicken Fen; and it might be thought that this would satisfy even the most academically minded biologist. But their mere occurrence here is not the point. They were exterminated before much was known about their original distribution, and it is only from the detailed distribution of freshwater animals that we can work out their past history. Much very impressive information on the glacial history of Europe has been built up in this way (Thienemann, 1950), and it is a sad thought that pollution should have removed some of the evidence before it was properly recorded. This is especially unfortunate because the species involved, being confined to the relatively stable habitat found in large rivers, are those most likely to have been affected by glacial conditions.

Finally I wish to stress once more the point, made in the first chapter, that the key to control of pollution is co-operation. Once Government, industry, riparian owners, fishermen, chemists, engineers and biologists are all aware of the problems involved, and when they are able to collaborate in a rational manner, the problem will not prove insoluble nor even very difficult to solve. It is, however, essentially a biological problem and its solution will not be possible without the help of biologists, who should play a much larger role in the future than they have in the past. Recrimination and litigation may sometimes seem to produce rapid results, but they do not lead to good feeling nor to collaboration between the various interests concerned with inland waters. Collaboration can only occur where there is mutual understanding, and I hope that this book will help to explain a biologist's point of view.

REFERENCES

Alabaster, J. S., Herbert, D. W. M., and Hemens, J. (1957). The survival of rainbow trout (*Salmo gairdneri* Richardson) and perch (*Perca fluviatilis* L.) at various concentrations of dissolved oxygen and carbon dioxide. *Ann. appl. Biol.*, **45**, 177–88.

Alexander, W. B., Southgate, B. A., and Bassindale, R. (1935). Survey of the river Tees, Part II. The estuary—chemical and biological. *Tech. Pap. Wat. Pollut. Res. Lond.*, **5**, 1–171.

Allan, I. R. H. (1955). Effects of pollution on fisheries. *Verh. int. Ver. Limnol.*, **12**, 804–10.

Allan, I. R. H., Herbert, D. W. M., and Alabaster, J. S. (1958.) A field and laboratory investigation of fish in a sewage effluent. *Fish. Invest. Lond.*, **1, 6, 2,** 76 pp.

Allen, L. A., Blezard, N., and Wheatland, A. B. (1946). Toxicity to fish of chlorinated sewage effluents. *Surveyor Lond.*, April 19th, 1 p.

Allen, L. A., Pasley, S. M., and Pierce, M. A. F. (1952). Some factors affecting the viability of faecal bacteria in water. *J. gen. Microbiol.*, **7**, 36–43.

American Public Health Assoc. (1946). *Standard methods for the examination of water and sewage*, 9th ed. New York.

Anderson, B. G. (1944). The toxicity thresholds of various substances found in industrial wastes as determined by the use of *Daphnia magna. Sewage Works Journ.*, **16,** 1156–65.

Anderson, B. G. (1946). The toxicity thresholds of various sodium salts determined by the use of *Daphnia magna. Sewage Works Journ.*, **18,** 82–7.

Anonymous (1948). Biological aspects of water pollution. *Nature Lond.*, **161,** 874–6.

Bachmann, H., Birrer, A., and Weber, H. (1934). Über die Giftwirkung von chemischen Substanzen auf niedere Wasserorganismen. *Schweiz Z. Hydrol.*, **6,** 63–127.

Bahr, H. (1953). Bakterielle Mikroorganismen im Rahmen der Eutrophie unserer Flüsse. *Zbl. Bakt.*, I, **160,** 85–9.

Bartsch, A. F. (1948). Biological aspects of stream pollution. *Sewage Works Journ.*, **20,** 292–302.

Berg, K. (1943). Physiographical studies on the river Susaa.*Folia limnol. scand.*, **1,** 170 pp.

Berg, K. (1948). Biological studies on the river Susaa. *Folia limnol. scand.*, **4**, 318 pp.

Berg, K. *et al.* (1958). *Furesøundersøgelser 1950–54. Folia limnol. scand.*, **10,** 189 pp.

Blum, J. L. (1957). An ecological study of the algae of the Saline river, Michigan. *Hydrobiologia*, **9**, 361–408.

REFERENCES

Braarud, T. (1955). The effect of pollution by sewage úpon the waters of Oslo-Fjord. *Verh. int. Ver. Limnol.*, **12,** 811–13.

Brinley, F. J. (1942). Biological studies, Ohio River Pollution Survey. *Sewage Works Journ.*, **14,** 147–59.

Burgess, S. G. (1957). The analysis of trade-waste waters. In Isaac 1957 (*op. cit.*), pp. 65–84.

Butcher, R. W. (1932a). Contribution to our knowledge of the ecology of sewage fungus. *Trans. Brit. mycol. Soc.*, **17,** 112–23.

Butcher, R. W. (1932b). Studies on the ecology of rivers, II. The micro-flora of rivers with special reference to the algae on the river bed. *Ann. Bot. Lond.*, **46,** 813–61.

Butcher, R. W. (1933). Studies on the ecology of rivers, I. On the distribution of macrophytic vegetation in the rivers of Britain. *J. Ecol.*, **21,** 58-91.

Butcher, R. W. (1946a). Studies on the ecology of rivers, VI. The algal growth in highly calcareous streams. *J. Ecol.*, **33,** 268–83.

Butcher, R. W. (1946b). The biological detection of pollution. *J. Inst. Sew. Purif.*, **2,** 92–7.

Butcher, R. W. (1947). Studies on the ecology of rivers, VII. The algae of organically enriched waters. *J. Ecol.*, **35,** 186–91.

Butcher, R. W. (1948). Problems of distribution of sessile algae in running water. *Verh. int. Ver. Limnol.*, **10,** 98–103.

Butcher, R. W. (1950). Survey of the river Derwent below Derby. *Rep. Trent Fish. Dist.*, 1950, 3–11.

Butcher, R. W. (1955). Relation between the biology and the polluted condition of the Trent. *Verh. int. Ver. Limnol.*, **12,** 823–7.

Butcher, R. W., Longwell, J., and Pentelow, F. T. K. (1937). Survey of the river Tees, III. The non-tidal reaches. Chemical and Biological. *Tech. Pap. Wat. Pollut. Res. Lond.*, **6,** 187 pp.

Butcher, R. W., Pentelow, F. T. K., and Woodley, J. W. A. (1927). The diurnal variation of the gaseous constituents of river waters. *Biochem. J.*, **21,** 945–57.

Butcher, R. W., Pentelow, F. T. K., and Woodley, J. W. A. (1931). An investigation of the river Lark and the effect of beet-sugar pollution. *Fish. Invest. Lond.*, **1, 3, 3,** 112 pp.

Campbell, M. S. A. (1939). Biological indicators of intensity of stream pollution. *Sewage Works Journ.*, **11,** 123–7.

Carpenter, K. E. (1924). A study of the fauna of rivers polluted by lead mining in the Aberystwyth district of Cardiganshire. *Ann. appl. Biol.*, **9,** 1–23.

Carpenter, K. E. (1925). On the biological factors involved in the destruction of river-fisheries by pollution due to lead-mining. *Ann. appl. Biol.*, **12,** 1–13.

Carpenter, K. E. (1926). The lead mine as an active agent in river pollution. *Ann. appl. Biol.*, **13,** 395–401.

Carpenter, K. E. (1927). The lethal action of soluble metallic salts on fishes. *Brit. J. exp. Biol.*, **4,** 378–90.

Carpenter, K. E. (1928). *Life in inland waters*. London.

Carpenter, K. E. (1930). Further researches on the action of metallic salts on fishes. *J. exp. Zool.*, **56**, 407–22.

Chang, S. H. (1949). Some epidemiological and biological problems in water-borne diseases. In Moulton and Hitzel 1949 (*op. cit.*), pp. 16–32.

Claassen, P. W. (1926). Biological studies of polluted areas in the Genesee river system. *Rep. N.Y. St. Conserv. Dep. Suppl.*, pp. 38–47.

Clemens, H. P., and Jones, W. H. (1954). Toxicity of brine water from oil wells. *Trans. Amer. Fish. Soc.*, **84**, 97–109.

Cole, A. E. (1941). The effects of pollutional wastes on fish life. *A symposium on hydrobiology*. Madison. pp. 241–59.

Collman, R. V. (1957). The impact of trade wastes on sewage-treatment practice. In Isaac 1957 (*op. cit.*) pp. 47–56.

Cooke, W. B. (1954). Fungi in polluted water and sewage. *Sewage industr. Wastes*, **26**, 539–49.

Davis, H. S. (1938). Instructions for conducting stream and lake surveys. *Fish. Circ.*, **26**, 55 pp.

Degens, Ir. P. N., Van Der Zee, H., Kommer, J. D., and Kamphius, A. H. (1950). Synthetic detergents and sewage processing, V. The effect of synthetic detergents on certain water fauna. *J. Inst. Sew. Purif.*, **1**, 63–8.

Department of Scientific and Industrial Research. (1938). The effect of the discharge of crude sewage into the estuary of the river Mersey on the amount and hardness of the deposit in the estuary. *Tech. Pap. Wat. Pollut. Res. Lond.*, **7**, 337 pp.

Domogalla, B. (1941). Scientific studies and chemical treatment of the Madison lakes. *A symposium on hydrobiology*. Madison, pp. 303–9.

Doudoroff, P., and Katz, M. (1950). Critical review of literature on the toxicity of industrial wastes and their components to fish, I. Alkalis, acids and inorganic gases. *Sewage industr. Wastes*, **22**, 1432–58.

Doudoroff, P., and Katz, M. (1953). Critical review of literature on the toxicity of industrial wastes and their components to fish, II. The metals, as salts. *Sewage industr. Wastes*, **25**, 802–39.

Downing, K. M., and Merkens, J. C. (1955). The influence of dissolved-oxygen concentration on the toxicity of un-ionized ammonia to rainbow trout (*Salmo gairdnerii* Richardson). *Ann. appl. Biol.*, **43**, 243–6.

Downing, K. M., and Merkens, J. C. (1957). The influence of temperature on the survival of several species of fish in low tensions of dissolved oxygen. *Ann. appl. Biol.*, **45**, 261–7.

Eastham, L. E. S. (1938). Movements of the gills of Ephemerid nymphs in relation to the water currents produced by them. *J. Queckett micr. Cl.*, (4) **1**, 95–9.

Eastham, L. E. S. (1939). Gill movements of nymphal *Ephemera danica* (Ephemeroptera) and the water currents caused by them. *J. exp. Biol.*, **16**, 18–33.

Edwards, R. W. (1957). Vernal sloughing of sludge deposits in a sewage effluent channel. *Nature Lond.*, **180**, 100.

Ellis, M. M. (1937). Detection and measurement of stream pollution. *Bull. U.S. Bur. Fish.*, **48**, 365–437.

Everhart, W. H., and Hassler, W. W. (1945). Aquarium studies on the toxicity of D.D.T. to brown trout, *Salmo trutta*. *Trans. Amer. Fish. Soc.*, **75**, 59–64.

Fitch, C. F., Bishop, L. M., Boyd, W. L., Gortner, R. A., Rogers, C. F., and Tilden, J. E. (1934). 'Water bloom' as a cause of poisoning in domestic animals. *Cornell Vet.*, **24**, 30–9.

Fjerdingstad, E. (1950). The microflora of the river Mølleaa with special reference to the relation of the benthal algae to pollution. *Folia limnol. scand.*, **5**, 123 pp.

Fjerdingstad, E. (1954). *Bodo minimus* Klein. Notes on its ecology and significance for estimation of sewage. *Hydrobiologia*, **6**, 328–30.

Flaigg, N. G., and Reid, G. W. (1954). Effects of nitrogenous compounds on stream conditions. *Sewage industr. Wastes*, **26**, 1145–54.

Gameson, A. L. H. (1957). Weirs and the aeration of rivers. *J. Instn. Wat. Engrs.*, **11**, 477–90.

Gameson, A. L. H., Truesdale, G. A., and Downing, A. L. (1955). Re-aeration studies in a lakeland beck. *J. Instn. Wat. Engrs.*, **9**, 571–94.

Gardiner, A. C. (1927). The effect of aqueous extracts of tar on developing trout ova, and on alevins. *Fish. Invest. Lond.*, **1, 3, 2,** 14 pp.

Gardner, J. A. (1926). Report on the respiratory exchange in freshwater fish, with suggestions as to further investigations. *Fish. Invest. Lond.*, **1, 3, 1,** 17 pp.

Garner, J. H., Brown, F. M., and Lovett, M. (1936). Chemical and biological survey of the river Holme. *W. Riding of Yorkshire Riv. Bd.*, Wakefield.

Gaufin, A. R., and Tarzwell, C. M. (1952). Aquatic invertebrates as indicators of stream pollution. *Publ. Hlth. Rep. Wash.*, **67, 1,** 57–64.

Gaufin, A. R., and Tarzwell, C. M. (1956). Aquatic macro-invertebrate communities as indicators of pollution in Lytle Creek. *Sewage industr. Wastes*, **28**, 906–24.

Gorham, E. (1958). The influence and importance of daily weather conditions in the supply of chloride, sulphate and other ions to fresh waters from atmospheric precipitation. *Phil. Trans. B*, **679**, 147–78.

Gray, E. A. (1956). The microbiology of a polluted stream. *Verh. int. Ver. Limnol.*, **12**, 814–17.

Greenbank, J. (1941). Selective poisoning of fish. *Trans. Amer. Fish. Soc.*, **70**, 80-6.

Grindley, J. (1946). Toxicity to rainbow trout and minnows of some substances known to be present in waste water discharged into rivers. *Ann. appl. Biol.*, **33**, 103–12.

Haempel, O. (1925). Die Einwirkung von bei der Papirfabrikation verwendten Farbstoffen auf die Tierwelt des Wassers. *Z. Untersuch. Nahr-. u. Genussm.*, **50**, 423–6.

Harrison, A. D. (1958a). The effects of sulphuric acid pollution on the biology of streams in the Transvaal. *Verh. int. Ver. Limnol.* **13,** 603–10.

Harrison, A. D. (1958b). The effects of organic pollution on the fauna of parts of the Great Berg River System and of the Krom Stream, Stellenbosch. *Trans. R. Soc. S. Africa,* **35,** 299–329.

Harvey, J. V. (1952). Relationship of aquatic fungi to water pollution. *Sewage industr. Wastes,* **24,** 1159–64.

Hasler, A. D. (1947). Eutrophication of lakes by domestic drainage. *Ecology,* **28,** 383–95.

Hawkes, H. A. (1957). Biological aspects of river pollution. In Klein 1957a (*op. cit.*), pp. 191–251.

Hentschel, E. (1925). Abwasserbiologie in Abderhalden E. *Handbuch der biologischen Arbeitsmethoden,* **9, 2, 1,** 233–80. Berlin.

Herbert, D. W. M. (1952). Measurement of the toxicity of substances to fish. *J. inst. Sew. Purif.*, Part 1, 60–8.

Herbert, D. W. M., Downing, K. M., and Merkens, J. C. (1955). Studies on the survival of fish in poisonous solutions. *Verh. int. Ver. Limnol.*, **12,** 789-94.

Herbert, D. W. M., Elkins, G. H. J., Mann, H. T., and Hemens, J. (1957). Toxicity of synthetic detergents to rainbow trout. *Wat. Waste Treatm. J.* Sept./Oct. 4 pp.

Herbert, D. W. M., and Mann, H. T. (1958). The tolerance of some fresh-water fish for sea water. *Salmon Trout Mag.*, May, 3 pp.

Herbert, D. W. M., and Merkens, J. C. (1952). The toxicity of potassium cyanide to trout. *J. exp. Biol.*, **29,** 632–49.

Hey, D. (1955). A preliminary report on the culture of fish in the final effluent from the new disposal works, Athlone, S. Africa. *Verh. int. Ver. Limnol.*, **12,** 737–42.

Hoffmann, C. H., and Drooz, A. T. (1953). Effects of a C 47 airplane application of D.D.T. on fish-food organisms in two Pennsylvania watersheds. *Amer. Midl. Nat.*, **50,** 172–88.

Hoffmann, C. H., and Surber, E. W. (1945). Effects of an aerial application of wettable D.D.T. on fish and fish-food organisms in Back Creek, West Virginia. *Trans. Amer. Fish. Soc.*, **75,** 48–58.

Huet, M. (1949). La pollution des eaux. L'Analyse biologique des eaux polluées. *Bull. Centr. Belg. Etude Doc. Eaux.*, **5** and **6,** 1–31.

Huet, M. (1950). Toxicologie des poissons. *Bull. Centr. Belg. Etude Doc. Eaux.*, **7,** 396–406.

Huet, M. (1951). Nocivité des boisements en Epicéas (*Picea excelsa* Link.) pour certains cours d'eaux de l'Ardenne Belge. *Verh. int. Ver. Limnol.*, **11,** 189–200.

Huet, M. (1954). Biologie, profils en long et en travers des eaux courantes. *Bull. franç. Piscic.*, **175,** 41–53.

Huet, M., Leclerc, E., Timmermans, J. A., and Beaujean, P. (1955). Recherche des corrélations entre l'analyse biologique et l'analyse physico-chimique des eaux polluées par matières organiques. *Bull. Centr. Belg. Etude Doc. Eaux.*, **30,** 216–37.

Huntsman, A. G. (1942). Death of salmon and trout with high temperature. *J. Fish. Res. Bd. Can.*, **5**, **5**, 485–501.

Hynes, H. B. N. (1941). The taxonomy and ecology of the nymphs of British Plecoptera with notes on the adults and eggs. *Trans. R. ent. Soc. Lond.*, **91**, 459–557.

Hynes, H. B. N. (1958a). The effect of drought on the fauna of a small mountain stream in Wales. *Verh. int. Ver. Limnol.*, **13**, 826–33.

Hynes, H. B. N. (1958b). A key to the adults and nymphs of British stoneflies. *Sci. Pub. Freshwat. biol. Ass.*, **17**, 86 pp.

Illies, J. (1955). Der biologische Aspekt der limnologischen Fleisswasser-typisierung. *Arch. Hydrobiol.* (*Plankt.*) *Suppl.*, **22**, 337–46.

Illies, J. (1956). Seeausflus-Biozönosen lapplandischer Waldbäche. *Ent. Tidskr.*, **77**, 138–53.

Imhoff, K. (1931). Impounding reservoirs as a substitute for biological sewage treatment works. *Sewage Works Journ.*, **3**, 120–4.

Imhoff, K. (1955). The final step in sewage treatment. *Sewage industr. Wastes*, **27**, 332–5.

Ingram, W. M., and Prescott, G. W. (1954). Toxic fresh-water algae. *Amer. Midl. Nat.*, **52**, 75–87.

Isaac, P. C. G. (ed.) (1957). The treatment of trade-waste waters and the prevention of river pollution. *Bulletin No.* 10, *Dept., of Civil Engineering, King's College, University of Durham.*

Isaac, P. C. G., and Lodge, M. (1958). Algae and sewage treatment. *New Biol.*, **25**, 85–97.

Jaag, O. (1955). Some effects of pollution on natural waters. *Verh. int. Ver. Limnol.*, **12**, 761–7.

Jenkins, S. H. (1957). The biological oxidation of trade wastes. In Isaac 1957 (*op. cit.*), pp. 115–42.

Jónasson, P. M. (1955). The efficiency of sieving techniques for sampling freshwater bottom fauna. *Oikos*, **6**, 183–207.

*Jones, J. R. E. (1937). The toxicity of dissolved metallic salts to *Polycelis nigra* (Muller), and *Gammarus pulex* (L.). *J. exp. Biol.*, **14**, 351–63.

Jones, J. R. E. (1938a). The relative toxicity of salts of lead, zinc and copper to the stickleback (*Gasterosteus aculeatus* L.) and the effect of calcium on the toxicity of lead and zinc salts. *J. exp. Biol.*, **15**, 934–407.

Jones, J. R. E. (1938b). Antagonism between two heavy metals in their toxic action on freshwater animals. *Proc. zool. Soc. Lond. A.*, **108**, 481–99.

Jones, J. R. E. (1939a). Antagonism between salts of the heavy and alkaline-earth metals in their toxic action on the tadpole of the toad *Bufo bufo bufo* (L.). *J. exp. Biol.*, **16**, 313–33.

Jones, J. R. E. (1939b). The relation between the electrolytic solution pressures of the metals and their toxicity to the stickleback (*Gasterosteus aculeatus* L.). *J. exp. Biol.*, **16**, 425–37.

* Jones, J. R. E. is referred to as Jones in the text: Jones, J. W. as Jones J. W.

Jones, J. R. E. (1940a). The fauna of the river Melindwr, a lead-polluted tributary of the river Rheidol in north Cardiganshire, Wales. *J. Anim. Ecol.*, **9**, 188–201.

Jones, J. R. E. (1940b). A study of the zinc-polluted river Ystwyth in north Cardiganshire, Wales. *Ann. appl. Biol.*, **27**, 368–78.

Jones, J. R. E. (1941a). The fauna of the river Dovey, West Wales. *J. Anim. Ecol.*, **10**, 12–24.

Jones, J. R. E. (1941b). A study of the relative toxicity of anions with *Polycelis nigra* as test animal. *J. exp. Biol.*, **18**, 170–81.

Jones, J. R. E. (1943). The fauna of the river Teifi, West Wales. *J. Anim. Ecol.*, **12**, 115–23.

Jones, J. R. E. (1947a). The oxygen consumption of *Gasterosteus aculeatus* L. in toxic solutions. *J. exp. Biol.*, **23**, 298–311.

Jones, J. R. E. (1947b). The reactions of *Pygosteus pungitius* L. to toxic solutions. *J. exp. Biol.*, **24**, 110–22.

Jones, J. R. E. (1948a). The fauna of four streams in the 'Black Mountain' district of South Wales. *J. Anim. Ecol.*, **17**, 51–65.

Jones, J. R. E. (1948b). A further study of the reaction of fish to toxic solutions. *J. exp. Biol.*, **25**, 22–34.

Jones, J. R. E. (1949a). A further study of calcareous streams in the 'Black Mountain' district of South Wales. *J. Anim. Ecol.*, **18**, 142–59.

Jones, J. R. E. (1949b). An ecological study of the river Rheidol: north Cardiganshire, Wales. *J. Anim. Ecol.*, **18**, 67–88.

Jones, J. R. E. (1950). A further ecological study of the river Rheidol, the food of the common insects of the main-stream. *J. Anim. Ecol.*, **19**, 159–74.

Jones, J. R. E. (1951a). The reactions of the minnow, *Phoxinus phoxinus* L., to solutions of phenol, ortho-cresol and para-cresol. *J. exp. Biol.*, **28**, 261–70.

Jones, J. R. E. (1951b.) An ecological study of the river Towy. *J. Anim. Ecol.*, **20**, 68–86.

Jones, J. R. E. (1952). The reactions of fish to water of low oxygen concentration. *J. exp. Biol.*, **29**, 403–15.

Jones, J. R. E. (1957). Fish and river pollution. In Klein 1957a (loc. cit.), pp. 159–90.

Jones, J. R. E. (1958). A further study of the zinc-polluted river Ystwyth. *J. Anim. Ecol.*, **27**, 1–14.

Jones, J. W. (1958). The spawning grounds of salmon. *Trout and Salmon* **35**, **3**, 5–6.

Jones, J. W. (1959). *The Salmon. New Nat.* London.

Jones, J. W., and Ball, J. N. (1954). The spawning behaviour of brown trout and salmon. *Brit. J. Anim. Behav.*, **2**, 103–14.

Kaiser, E. W. (1951). Biologiske, biokemiske, bacteriologiske samt hydrometriske undersøgelser af Poleten 1946 og 1947. *Dansk. Ingenforen. Skr.*, **3**, 15–33.

Kaplovsky, A. J., and Harmic, J. L. (1953). Pollution study of the Red Clay Creek drainage basin. *Sewage industr. Wastes*, **25**, 1072–6.

Kerswill, C. J., and Elson, P. F. (1955). Preliminary observations on effects of D.D.T. spraying on Miramichi salmon stocks. *Progr. Rep. Atl. biol. Sta.*, **62**, 17–23.

Kisskalt, K., and Ilzhöffer (1937). Die Reinigung von Abwasser in Fischteichen. *Arch. Hyg. Berl.*, **118**, 1–66.

Klein, L. (1957a). *Aspects of river pollution*. London.

Klein, L. (1957b). Activated sludge for the treatment of trade wastes. In Isaac 1957 (*op. cit.*), pp. 143–58.

Klingler, K. (1957). Natriumnitrit, ein langsamwirkendes Fischgift. *Schweiz. Z. Hydrol.*, **19**, 565–78.

Kolkwitz, R. (1950). Oekologie der Saprobien. Über die Beziehungen der Wasserorganismen zur Umwelt. *SchrReihe Ver. Wasserhyg.*, **4**, 64 pp.

Kolkwitz, R., and Marsson, M. (1908). Ökologie der pflanzlichen Saprobien. *Ber. dtsch. bot. Ges.*, **26**, 505–19.

Kolkwitz, R., and Marsson, M. (1909). Ökologie der tierische Saprobien. Beiträge zur Lehre von der biologische Gewässerbeurteilung. *Int. Rev. Hydrobiol.*, **2**, 126–52.

Lackey, J. B. (1938). Protozoan plankton as indicators of pollution in a flowing stream. *Publ. Hlth. Rep. Wash.*, **53**, 2037–58.

Lackey, J. B. (1941). The significance of plankton in relation to the sanitary condition of streams. *A symposium on hydrobiology*. Madison, pp. 311–28.

Lackey, J. B. (1945). Plankton productivity of certain south-eastern Wisconsin lakes as related to fertilisation, II. Productivity. *Sewage Works Journ.*, **17**, 795–802.

Lackey, J. B. (1949). Plankton as related to nuisance conditions in surface water. In Moulton and Hitzel 1949 (*op. cit.*), pp. 56–63.

Lackey, J. B., and Sawyer, C. N. (1945). Plankton productivity of certain south-eastern Wisconsin lakes as related to fertilisation, I. Surveys. *Sewage Works Journ.*, **17**, 573–85.

Lackey, J. B., and Wattie, E. (1940). Studies of sewage purification, XIII. The biology of *Sphaerotilus natans* Kutzing in relation to the bulking of activated sludge. *Sewage Works Journ.*, **12**, 669–84.

Lafleur, R. A. (1954). Biological indices of pollution as observed in Louisiana streams. *Bill. La. Engng. exp. Sta.*, **43**, 1–7.

Larsen, K., and Olsen, S. (1950). Ochre suffocation of fish in the river Tim Aa. *Rep. Danish biol. Sta.*, **50**, 5–27.

Laurent, P. J. (1958). Résistance de crustacés d'eau douce à l'action du chromate de sodium. *Verh. int. Ver. Limnol.*, **13**, 590–5.

Laurie, R. D., and Jones, J. R. E. (1938). The faunistic recovery of a lead-polluted river in north Cardiganshire, Wales. *J. Anim. Ecol.*, **7**, 272–89.

Lea, W. L. (1941). The role of nitrogen and phosphorus in biochemical oxygen demand dilution water. *A symposium on hydrobiology*. Madison, pp. 71–85.

Lea, W. L., Rohlich, C. A., and Katz, W. J. (1954). Removal of phosphates from treated sewage. *Sewage industr. Wastes*, **26**, 261–75.

Leonard, J. W. (1939). Notes on the use of derris as a fish poison. *Trans. Amer. Fish. Soc.*, **68**, 269–79.

Liebmann, H. (1942). Die Bedeutung der mikroskopischen Untersuchung für die biologische Wasseranalyse. *Vom Wasser*, **15**, 181–8.

Liebmann, H. (1951). *Handbuch der Frischwasser und Abwasserbiologie.* Munich.

Liebmann, H. (1953). The inhabitants of *Sphaerotilus* flocs and the physical chemical conditions of their formation. *Vom Wasser.*, **20**, 24–33. (*Wat. Pollut. Abstr.*, **28**, 940).

Liepolt, R. (1953). Lebensraum und Lebensgemeinschaft des Liesingbaches. *Wett. u. Leben.*, *Sonderheft*, **2**, 64–102.

Liepolt, R. (1958). Gewässerverunreiningung in Österreich durch Holzindustrieabwässer. *Verh. int. Ver. Limnol.*, **13**, 481–90.

Lindroth, A. (1949). Vitality of salmon parr at low oxygen pressure. *Rep. Inst. Freshw. Res. Drottning.*, **29**, 49–50.

Lloyd, L. L. (1944). Sewage bacteria bed fauna in its natural setting. *Nature Lond.*, **154**, 397.

Lloyd, L. L. (1947). The biology of sewage disposal. *New Biol.*, **2**, 30–52.

Lovett, M. (1957). River pollution—General and chemical effects. In Isaac 1957 (*op. cit.*), pp. 9–26.

Lowndes, A. G. (1952). Hydrogen ion concentration and the distribution of freshwater Entomostraca. *Ann. Mag. nat. Hist.*, (12), **5**, 58–65.

Ludwig, H. F., and Oswald, W. J. (1952). Role of algae in sewage oxidation ponds. *Sci. Mon., N.Y.*, **74**, **1**, 3–6.

Macan, T. T. (1958). Methods of sampling the bottom in stony streams. *Mitt. int. Ver. Limnol.*, **8**, 21 pp.

Macan, T. T., Mortimer, C. H., and Worthington, E. B. (1942). The production of freshwater fish for food. *Sci. Pub. Freshwat. biol. Ass.*, **6**, 36 pp.

Macan, T. T., and Worthington, E. B. (1951). *Life in lakes and rivers. New Nat.*, **15**, London.

Mack, B. (1953). Zur Algen und Pilzflora des Liesingbaches. *Wett. u. Leben. Sonderheft*, **2**, 136–49.

Mackenthun, K. M. (1950). Aquatic weed control with sodium arsenite. *Sewage industr. Wastes*, **22**, 1062–7.

Mackenthun, K. M., Herman, E. F., and Bartsch, A. F. (1945). A heavy mortality of fishes resulting from the decomposition of algae in the Yahara river, Wisconsin. *Trans. Amer. Fish. Soc.*, **75**, 175–80.

Mallman, W. L. (1940). A new yardstick for measuring sewage pollution. *Sewage Works Journ.*, **12**, 875–8.

Mann, K. H. (1952). A revision of the British leeches of the family Erpobdellidae, with a description of *Dina liniata* (O. F. Müller 1774) a leech new to British fauna. *Proc. zool. Soc. Lond.*, **122**, 395–405.

Mann, K. H. (1953). A revision of the British leeches of the family Glossiphoniidae, with a description of *Batracobdella paludosa* (Carena 1824) a leech new to the British fauna. *Proc. zool. Soc. Lond.*, **123**, 377–91.

Mann, K. H. (1958). Occurrence of an exotic oligochaete *Branchiura sowerbyi* Beddard, 1892, in the River Thames. *Nature Lond.*, **182**, 732.

Margalef, R. (1949). A new limnological method for the investigation of thin-layered epilithic communities. *Hydrobiologia*, **1**, 215–16.

Merkens, J. C., and Downing, K. M. (1957). The effect of tension of dissolved oxygen on the toxicity of un-ionised ammonia to several species of fish. *Ann. appl. Biol.*, **45**, 521–7.

Meuwis, A. L., and Heuts, M. J. (1957). Temperature dependence of breathing rate of carp. *Biol. Bull., Wood's Hole*, **112**, 97–107.

Ministry of Health (1949). *Prevention of river pollution.* H.M.S.O., London.

Ministry of Housing and Local Government (1956). *Report of the Committee on Synthetic Detergents.* H.M.S.O., London.

Mohr, J. L. (1952). Protozoa as indicators of pollution. *Sci. Mon., N.Y.*, **74**, **1**, 7–9.

Moon, H. P. (1935). Flood movements of the littoral fauna of Windermere. *J. Anim. Ecol.*, **4**, 216–28.

Moon, H. P. (1939). Aspects of the ecology of aquatic insects. *Trans. Soc. Brit. Ent.*, **6**, 39–49.

Mossewitsch, N. A., and Gussew, A. G. (1958). Der Einfluss der Abwässer von Holzverarbeitungswerken auf Gewässer und Wasserorganismen. *Verh. int. Ver. Limnol.*, **13**, 525–32.

Mottley, C. McC., Rayner, H. J., and Rainwater, J. H. (1939). The determination of the food grade of streams. *Trans. Amer. Fish. Soc.*, **68**, 336–43.

Moulton. F. R., and Hitzel, F. (ed.) (1949). *Limnological aspects of water supply and waste disposal. Amer. Ass. Adv. Sci.* Washington, D.C.

Moyle, J. B. (1949). The use of copper sulphate for algal control and its biological implications. In Moulton and Hitzel 1949 (*op. cit.*)

Mueller, P. K. (1954). Effect of phenols on micro-organisms in the river Pleisse. *Wasserwirtschaft, Wien*, **4**, 125. (*Abs. Sew. Industr. Wastes*, 1954, **26**, 1510–11).

Müller, K. (1959). Die Einflusse der Flösserei auf die Fischerei in Schwedish-Lappland. *Verh. int. Ver. Limnol.*, **13**, 533–44.

Naylor, E. (1958). The fauna of a warm dock. *Read to XV Int. Congr. Zool., Sect.* **3**, *Paper* 12, 3 pp.

Newton, L. (1944). Pollution of the rivers of West Wales by lead and zinc mine effluent. *Ann. appl. Biol.*, **31**, 1–11.

Nitardy, E. (1942). Der Selbstreinigungsprozess des Rohabwassers im Laboratoriumsversuch (Vergleich der biologischen und chemisch-physikalischen Befunde). *Kleine Mitt. Ver. Wasser- u. Lufthyg.*, **18**, 13–41.

Nowak, W. (1940). Über der Verunreinigung eines kleinen Flusses in Mähren durch Abwasser von Weissgerbereien, Leder-, Leimfabriken und anderen Betrieben. *Arch Hydrobiol.* (*Plankt.*), **36**, 386–423.

Ohle, W. (1955). Die Ursachen der rasenten Seeneutrophierung. *Verh. int. Ver. Limnol.*, **13**, 373–82.

Olszewski, W., and Spitta, O. (1931). Ohlmüller-Spitta. *Untersuchung und Beurteilung des Wassers und des Abwassers.* Berlin.

Owen, R. (1953). Removal of phosphorus from sewage plant effluent with lime. *Sewage industr. Wastes*, **25**, 548–56.

Painter, H. A. (1954). Factors effecting the growth of some fungi associated with sewage purification. *J. gen. Microbiol.*, **10**, 177–90.

Patrick, R. (1950). Biological measure of stream conditions. *Sewage industr. Wastes*, **22**, 926–38.

Patrick, R. (1951). A proposed biological measure of stream conditions. *Verh. int. Ver. Limnol.*, **11**, 299–307.

Patrick, R. (1953). Aquatic organisms as an aid in solving waste disposal problems. *Sewage industr. Wastes*, **25**, 210–4.

Patrick, R. (1954). Diatoms as an indication of river change. Proc. 9th Industr. Waste Conf. *Purdue Univ. Engng. Extn. Ser.*, **87**, 325–30.

Paul, R. M. (1952). Water pollution: a factor modifying fish populations in Pacific Coast streams. *Sci. Mon., N.Y.*, **74**, **1**, 14–17.

Pearsall, W. H. (1949). The English lakes and their development. *New Biol.*, **6**, 9–28.

Pearsall, W. H., Gardiner, A. C., and Greenshields, F. (1946). Freshwater biology and water supply in Britain. *Sci. Pub. Freshwat. biol. Ass.*, **11**, 90 pp.

Pearsall, W. H., and Pennington, W. (1947). Ecological history of the English Lake District. *J. Ecol.*, **34**, 137–48.

Pehrson, S. O. (1958). Stream improvement by better technics in the pulp and paper industry. *Verh. int. Ver. Limnol.*, **13**, 455–62.

Pentelow, F. T. K. (1949). Fisheries and pollution from china clay works. *Rep. Salm. Freshw. Fish. Lond.*, **31**, 4 pp.

Pentelow, F. T. K. (1953). *River Purification.* London.

Pentelow, F. T. K. (1955). Pollution and Fisheries. *Verh. int. Ver. Limnol.*, **12**, 768–71.

Pentelow, F. T. K. (1958). River Boards in Britain. *Nature Lond.*, **181**, 1568–9.

Pentelow, F. T. K., and Butcher, R. W. (1938). Observations on the condition of Rivers Churnet and Dove in 1938. *Rep. Trent Fish. Dist. App. 1.*

Pentelow, F. T. K., Butcher, R. W., and Grindley, J. (1938). An investigation of the effects of milk wastes on the Bristol Avon. *Fish. Invest. Lond.*, **1**, **4**, **1**, 80 pp.

Percival, E., and Whitehead, H. (1929). A quantitative study of the fauna of some types of stream-bed. *J. Ecol.*, **17**, 282–314.

Percival, E., and Whitehead, H. (1930). Biological survey of the river Wharfe, II. Report on the invertebrate fauna. *J. Ecol.*, **18**, 286–302.

Phelps, E. B. (1944). *Stream sanitation.* New York.

Pielou, D. P. (1946). Lethal effects of D.D.T. on young fish. *Nature Lond.*, **158**, 378 pp.

Pirie, N. W. (1958). Unconventional production of foodstuffs. *Sci. News*, **49**, 17–38.

Postgate, J. (1954). The sulphur bacteria. *New Biol.*, **17**, 58–76.

Prescott, G. W. (1939). Some relationships of phytoplankton to limnology and aquatic biology. *Problems of lake biology. Amer. Ass. Adv. Sci. Pub.* No. 10, 65–78.

Prescott, G. W. (1948). Objectionable algae with reference to the killing of fish and other animals. *Hydrobiologia*, **1**, 1–13.

Pringsheim, E. G. (1949). The filamentous bacteria *Sphaerotilus*, *Leptothrix*, *Cladothrix*, and their relation to iron and manganese. *Phil. Trans. B.*, **233**, 453–82.

Pruthi, H. S. (1927). Preliminary observations on the relative importance of the various factors responsible for the death of fishes in polluted waters. *J. Mar. biol. Ass. U.K.*, **14**, 729–37.

Rasmussen, C. J. (1955). On the effect of silage juice in Danish streams. *Verh. int. Ver. Limnol.*, **12**, 819–22.

Redeke, H. C. (1927). Report on the pollution of rivers and its relation to fisheries. *Rapp. Cons. Explor. Mer.*, **43**, 1–50.

Rees, W. J. (1954). Biology of pollution. *J. Inst. Biol.*, **1**, 29–30.

Reese, M. J. (1937). The microflora of the non-calcareous streams Rheidol and Melindwr with special reference to water pollution from lead mines in Cardiganshire. *J. Ecol.*, **25**, 385–407.

Rennerfelt, J. G. V. (1958). B.O.D. of pulp wastes, its determination and importance. *Verh. int. Ver. Limnol.*, **13**, 545–56.

Richardson, R. E. (1921). Changes in the bottom and shore fauna of the middle Illinois River and its connecting lakes since 1913–15 as a result of increase southward of sewage pollution. *Bull. Ill. nat. Hist. Surv.*, **14**, 33–75.

Richardson, R. E. (1929). The bottom fauna of the middle Illinois River, 1913–1925: its distribution, abundance, valuation and index value in the study of stream pollution. *Bull. Ill. nat. Hist. Surv.*, **17**, 387–475.

Ricker, W. E. (1937). An ecological classification of certain Ontario streams. *Univ. Toronto Stud. biol.*, **37**, 114 pp.

Roberts, C. H., Grindley, J., and Williams, E. H. (1940). Chemical methods for the study of river pollution. *Fish. Invest. Lond.*, **1**, **4**, **2**, 34 pp.

Rudolfs, W. (1941). The microbiology of sewage and sewage treatment. *A symposium on hydrobiology*. Madison, pp. 273–9.

Schmitz, W. (1957). Die Bergbach-Zoozönosen und ihre Abgrenzung, dargestellt am Beispiel der oberen Fulda. *Arch. Hydrobiol.* (*Plankt.*), **53**, 465–98.

Schmitz, W. (1958). Ökologisch-physiologische Probleme der Besiedlung versalzener Binnengewässer. *Verh. int. Ver. Limnol.*, **13**, 959–60.

Schneller, M. V. (1955). Oxygen depletion in Salt Creek, Indiana. *Invest. Ind. Lakes*, **4**, 163–75.

Silverman, P. H. (1955). The Biology of Sewers and Sewage Treatment. The survival of the egg of the 'Beef Tapeworm', *Taenia saginata*. *Advance. Sci. Lond.*, **12**, 108–11.

Silverman, P. H., and Griffiths, R. B. (1955). A review of methods of sewage disposal in Great Britain, with special reference to the epizootiology of *Cysticercus bovis*. *Ann. trop. Med. Parasit.*, **49**, 436–50.

Simmonds, I. G. (1952). Investigation of wood fibre discharges into Spanish river. *Munic. Util.*, **12**, 19–24.

Sládeček, V. (1958). Die Abhängigkeit de Belebtschlammverfahrens von physikalischen, chemischen und biologischen Faktoren. *Verh. int. Ver. Limnol.*, **13**, 611–16.

Southgate, B. A. (1932). The toxicity of mixtures of poisons. *Quart. J. Pharm.*, **5**, **4**, 639–48.

Southgate, B. A. (1948). *Treatment and disposal of industrial waste waters.* H.M.S.O., London.

Southgate, B. A. (1951). Pollution of streams: some notes on recent research. *Inst. Civil Engineers. Public Health Paper No. 1.*

Southgate, B. A. (1957a). Synthetic detergents—a new pollution problem. *J. R. Soc. Arts*, **55**, 485–97.

Southgate, B. A. (1957b). A survey of trade-waste treatment. In Isaac 1957 (*op. cit.*), pp. 57–64.

Southgate, B. A., Pentelow, F. T. K., and Bassindale, R. (1933). The toxicity to trout of potassium cyanide and *p*-cresol in water containing different concentrations of dissolved oxygen. *Biochem. J.*, **27**, 983–5.

Šrámek-Hušek, R. (1958). Die Rolle der Ciliatenanalyse bei der biologischen Kontrolle von Flussverunreinigungen. *Verh. int. Ver. Limnol.*, **13**, 636–45.

Stammer, H. A. (1953). Der Einfluss von Schwefelwasserstoff und Ammoniak auf tierische Leitformen des Saprobiensystems. *Vom Wasser*, **20**, 34–71.

Steeman Nielsen, E. (1955). The production of organic matter by the phytoplankton in a Danish lake receiving extraordinarily great amounts of nutrient salts. *Hydrobiologia*, **7**, 68–74.

Steinmann, P., and Surbeck, G. (1918). *Die Wirkung organischer Verunreinigungen auf die Fauna schweizerischer fleissender Gewässer.* Bern.

Steinmann, P., and Surbeck, G. (1922). Zum Problem der biologischen Abwasseranalyse. *Arch. Hydrobiol.* (*Plankt.*), **13**, 404–12.

Stjerna-Pooth, I. (1957). *Achlya prolifera* als Abwasserpilz in einem mittelschwedischen Wasserlauf. *Rep. Inst. Freshw. Res. Drottning.*, **38**, 247–66.

Stopford, S. C. (1951). An ecological survey of the Cheshire foreshore of the Dee estuary. *J. Anim. Ecol.*, **20**, 103–22.

Stuart, T. A. (1953). Water currents through permeable gravels and their significance to spawning Salmonids, etc., *Nature Lond.*, **172**, 407.

Stundl, K. (1958). Versuche über die Wirkung von Abwässern holzverarbeitender Industrien auf Vorflutorganismen. *Verh. int. Ver. Limnol.*, **13**, 507–13.

Suckling, E. V. (1944). *The examination of waters and water supplies.* 5th ed. London.

Surber, E. W. (1953). Biological effects of pollution in Michigan waters. *Sew. industr. Wastes*, **25**, 79–86.

Tate Regan, C. (1911). *The freshwater fishes of the British Isles.* London.

Thienemann, A. (1950). Verbreitungsgeschichte der Süsswasserfauna Europas. *Binnengewässer*, **18.**

Thienemann, A. (1954). Chironomus. *Binnengewässer*, **20.**

Thomas, E. A. (1944). Versuche über die Selbstreinigung fliessenden Wassers. *Mitt. Lebensm. Hyg. Bern.*, **35**, 199–216.

Thomas, E. A. (1953). Zur Bekämpfung der See-Eutrophierung. *Monatsbull. schweiz. Ver. Gas- u. Wasserfachm.*, **2/3**, 1–15.

Thomas, E. A. (1955a). Über die Bedeutung der abwasserbedingten direkten Sauerstoffzehrung in Seen. *Monatsbull. schweiz. Ver. Gas- u. Wasserfachm.*, **5**, 1–11.

Thomas, E. A. (1955b). Phosphateghalt der Gewässer und Gewässerschutz. *Monatsbull. schweiz. Ver. Gas- u. Wasserfachm.*, **9/10**, 1–16.

Thomas, E. A. (1956–7). Der Zürichsee, sein Wasser und sein Boden. *Jahrbuch vom Zürichsee*, **17**, 173–208.

Tomlinson, T. G. (1945). Control by D.D.T. of flies breeding in percolating sewage filters. *Nature Lond.*, **156**, 478.

Tomlinson, T. G., Grindley, J., Collet, R., and Muirden, M. J. (1949). Control of flies breeding in percolating sewage filters (Part 2.) *J. Inst. Sewage Purif.*, Part 2, 127–9.

Tomlinson, T. G., and Muirden, M. J. (1948). Control of flies breeding in percolating sewage filters. *J. Inst. Sewage Purif.*, Part 1, 127–39.

Turing, H. D. (1947–49). Four reports on pollution affecting rivers in England, Wales and Scotland. British Field Sports Society. London.

Turoboyski, L. (1953). Infusoria as indicators of pollution in the Vistula river below Cracow. *Gaz. Woda Tech. Sanit.*, **27**, 326. (*Abs. Sew. industr. Wastes*, **26**, 1055.)

United States Department of the Interior (1957). Preliminary toxicity studies with hexadecanol reservoir evaporation reduction. *Bureau of Reclamation Chemical Engineering Lab. Rep.*, No. 51–10.

Vallin, S. (1958). Einfluss der Abwässer der Holzindustrie auf den Vorfluter. *Verh. int. Ver. Limnol.*, **13**, 463–73.

Van Horn (1949). A study of kraft pulping wastes in relation to the aquatic environment. In Moulton and Hitzel 1949 (*op. cit.*), pp. 49–55.

Van Oosten, J. (1945). Turbidity as a factor in the decline of Great Lakes fishes with special reference to Lake Erie. *Trans. Amer. Fish. Soc.*, **75**, 281–322.

Wallen, I. E. (1951). The direct effect of turbidity on fishes. *Bull. Okla. agric. mech. Coll., Biol. Ser.*, **48**, **2**, 27 pp.

Walshe, B. M. (1950). Observations on the biology and behaviour of larvae of the midge *Rheotanytarsus*. *J. Queckett. micr. Cl. ser 4*, **3**, 171–8.

Walshe, B. M. (1951). The feeding habits of certain chironomid larvae (subfamily Tendipedinae). *Proc. zool. Soc. Lond.*, **121**, 63–79.

Wang, W. L. L., and Dunlop, S. G. (1954). Animal parasites in sewage and irrigation water. *Sewage industr. Wastes*, **26**, 1020–32.

Welch, P. S. (1935). *Limnology*. New York.

Wesenberg-Lund, C. (1939). *Biologie der Süsswassertiere.* Vienna.

Wesenberg-Lund, C. (1943). *Biologie der Süsswasserinsekten.* Copenhagen.

Westlake, D. F., and Edwards, R. W. (1956). Director's Report. *Rep. Freshw. biol. Ass. Brit. Emp.*, **24**, 37–9.

Westlake, D. F., and Edwards, R. W. (1957). Director's Report. *Rep. Freshw. biol. Ass. Brit. Emp.*, **25**, 35–7.

Whipple, G. C. (1947). *The microscopy of drinking water.* 4th ed. revised by Fair, G. M. and Whipple, M. C. New York.

Whitehead, H. (1935). An ecological study of the invertebrate fauna of a chalk stream near Great Driffield, Yorkshire. *J. Anim. Ecol.*, **4**, 58–78.

Wiebe, A. H. (1927). Biological survey of the upper Mississippi river with special reference to pollution. *Bull. U.S. Bur. Fish.*, **43**, 137–67.

Wilson, J. N. (1949). Microbiota of sewage treatment plants and polluted streams. In Moulton and Hitzel 1949 (*op. cit.*), pp. 1–15.

Wilson, J. N. (1953). Effect of kraft mill wastes on stream bottom fauna. *Sewage industr. Wastes*, **25**, 1210–18.

Wilson, W. L. (1957). The treatment, conveyance and disposal of radioactive wastes. In Isaac 1957 (*op. cit.*), pp. 253–72.

Wisdom, A. S. (1956). *The law on the pollution of waters.* London.

Wuhrmann, K., and Woker, H. (1948). Experimentalle Untersuchungen über die Ammoniak- und Blausäurevergiftung. *Schweiz Z. Hydrol.*, **11**, 210–44.

Wuhrmann, K., and Woker, H. (1955). Influence of temperature and oxygen tension on the toxicity of poisons to fish. *Verh. int. Ver. Limnol.*, **12**, 795–801.

Wuhrmann, K., and Woker, H. (1958). Vergiftungen der aquatischen Fauna durch Gewässerverunreiningungen. *Verh. int. Ver. Limnol.*, **13**, 557–83.

Wuhrmann, K., Zehender, F., and Woker, H. (1947). Über die Fischereibiologische Bedeutung des Ammonium- und Ammoniakgehaltes fliessender Gewässer. *Vjschr. naturf. Ges. Zürich*, **42**, 198–204.

Wurtz, C. B. (1955). Stream biota and stream pollution. *Sewage industr. Wastes*, **27**, 1270–8.

Young, L. A., and Nicholson, H. P. (1951). Stream pollution resulting from the use of organic insecticides. *Progr. Fish. Cult.*, **13**, **4**, 193. (*Abs. Sew. industr. Wastes*, 1952, **24**, 1056.)

SUBJECT INDEX

Page numbers in bold type refer to illustrations of organisms

A

Aberystwyth, University College, 79
Achlya, 100
Achnanthes, 24, **25**, 81
acidity, See pH
acids, 53, 55, 74–5
activated sludge, 56–7, 109
Acts of Parliament, 3–6
adaptations of stream animals, 32–5
Agapetus, **31**, 42, 44–5, 123–4, 128–30
agriculture, 1, 53, 150
Agrion, 40
alcohol, 78
alderfly, see *Sialis*
Aldrin, 150
alevins, 74
algae,
  blooms of, 11, 141, 144, 152, 171
  control of, 152–3, 171
  effects of poisons on, 80–2
  effects of suspended solids on, 88
  as food of invertebrates, 35, 80
  growth of on glass slides, 24, 80
  of lake shores, 144
  of organically polluted water, 94, 101–6, 122
  as oxygenators, 23
  of plankton, 11–**12**, 26, 141–2
  as poisons, 141, 145
  in purification plant, 56, 172
  of rivers, 23–6
algicides, 152
alkalis, 53, 55, 75
alkyl aryl sulphonates, 67, 75, 169
alkyl sulphates, 67
Alps, 14
ammonia, 53, 58, 77, 82, 84, 92, 114, 119–20
  ionisation of, 58, 75
  oxidation of, 58, 62
  production of, 64–5
  toxicity of, 75, 120
Amoeba, 111
Amphinemura, **30**, 40, 44, 47–9, 82–3, 91, 123, 126, 128–30
Amphipoda, see shrimps.
Anabaena, **12**, 141, 144
Anabolia, 38, **39**
analytical techniques (chemical), 57–60
Ancylastrum (limpet), 28, **31**–2, 38, 44–6, 50, 79, 90, 123, 125, 127, 130
angiosperms, see weeds.
Ankistrodesmus, **12**
Anodonta, 36
Apium (water parsnip), 23, 40
Apodya, **99**
Arenicola, 132
Aristotle, 2
arsenic, 71, 152
Artemia, 140
Ascaris, 134
Asellus (water louse), 36, **37**, 45–6, 81, 93–4, 113, 115–8, 129–30, 140, 149
Asterionella, **12**, 144
atomic power, 137, 173

B

bacteria, 54–5, 58, 65, 93–5, 138
  anaerobic, 65–6, 95
  as constituents of sewage fungus, 95–8
  in cooling systems, 54, 138
  estimation of, 135, 155
  iron, 54, 86, 96
  in mud, 95
  in organic effluents, 93–5
  as pathogens, 95, 133–5
  in pipes, 57
  sulphur, 98, 133
Bacterium coli, see *Escherischia*
Baetis, **29**, 32, 38, 40, 42, 44–9, 51, 83–4, 89, 123, 125–30
barnacle, 137
Batrachospermum, 79
bean shrimps, see Ostracoda
beetles (Coleoptera), **30**, 35, 41, 44–6, 48, 83, 90, 123, 125, 127–8, 140, 147, 151
Beggiatoa, 56, 98, **99**, 132–3, 166

Berula (water celery), 22
B.H.C., 58, 150–1
biochemical oxygen demand, 4, 58, 89, 106, 138–9, 155–7, 169, 172
  measurement of, 59, 62
  significance of, 59–60
biological treatment of effluents, 55–7 61–3, 133, 170, 172
birds feeding on sewage, 134
Bithynia, 36, **37**
bivalve molluscs, 36, **37, 39,** 78, 89, 90, 132
blackening of mud and stones, 62, 95, 133
blanket weed, see *Cladophora*
blood worms, see *Chironomus*
B.O.D., see biochemical oxygen demand
Bodo, 109, **110,** 159, 164
Brachyptera, 48–9
brackish water, 97–8, 131–3, 140, 144–5
Branchiura, 137
breweries, 54, 68, 98
brine, 67, 139–40
brine shrimp, see *Artemia*
Bristol, 3
buffalo gnat, see *Simulium*
bugs (Hemiptera), 46, 148
bull rush, see *Schoenoplectus*
burr reed, see *Sparganium*

C

caddis-worms and flies, 27, **31**–3, 35, 38, **39,** 40, 44–9, 76, 80, 83–4, 89–91, 123–30, 140, 147, 149, 151
Caenis, **37,** 38, 44, 46, 91
calcium, 18, 28, 43, 75–6, 140 and see hardness.
Callitriche (starwort), 22–3, 28, 79, 107
canadian pond-weed, see *Elodea*
canalisation of rivers, 147–8
canals, 131, 147
canning plants, 54
carbohydrates, 56, 97
carbon dioxide, 15, 18, 53, 56, 62, 119–20, 172
Carchesium, **99, 100,** 109
Carcinus (crab), 132
Cardium (cockle), 132
Carex (sedge), 23
cattle yards, 54
cellulose factories, 98
Central Advisory Water Committee, 5
Centroptilum, 40, **41**
Ceratium, **12**
Ceratoneis, 24
Ceratopogonidae (biting midges), **33,** 45–6, 83–4, 125, 140
cetyl alcohol, 153
Chaetophorales, 24
Chamaesiphon, **25,** 81, 103–4, 122
Chilodonella, 109, **110**
china clay, 53, 87
Chironomidae, 27, 32, **33–7,** 40, **41**–6, 48–9, 71–2, 81–4, 89, 90, 113–8, 123–5, 129–30, 148, 151, 154, and see Orthocladiinae, Tanypodinae and *Tanytarsus*.
Chironomus (blood worm), 36, **37,** 42, 78, 81, 89, 93–4, 113–8, 121, 130, 140, 165
  as test for insecticides, 150
Chlamydomonas, **12**
chlorination, 54, 134, 138, 144
chlorine, 54, 58, 76, 134, 138
Chlorococcum, 81
chloroform, 78
Chloroperla, **30,** 32, 44, 46, 48–9, 128
cholera, 3, 133
chromium, 53
Cladophora (blanket weed), **25,** 82, 93–4, 105–7, 115–7, 124, 140
Cladothrix, 96, 98
clam, see *Mya*
Clitellio, 132
Closterium, 105
coal forests, 1
coal washing, 53, 88
Coccolithus, 145
Cocconeis, **25,** 81, 103–4, 122
cockle, see *Cardium*
Coleps, 109, **110**
Colpidium, 109, **110,** 160
Common Law, 5
common reed, see *Phragmites*
copper, 53, 72–6, 78, 80–1, 130, 152
copper sulphate as algicide, 152
Coregonus, see fish species.
Corixa, **41**–2
Cornwall, 2
Corophium, 132
Cosmarium, **12,** 144
crabs, 132, 137
craneflies, see Tipulidae and *Dicranota*
Crangon, 132
Crangonyx, 127–8, 147
crayfish, 78, 139
Crenobia, **29,** 33

cresols, 75, and see phenols
Cricotopus, **41**, 115
Crucigenia, **12**
Crunoecia, 44
Culex (mosquito), 111, **112**, 165
Culicoides, 140
cultivation, see agriculture
cyanide, 53, 58, 75, 82, 84, 92, 119, 130
cyanogen chloride, 74, 138
Cyclops, **39**
Cymbella, **25**
Cypris, 36, **37**

D

dairies, 54, 97, and see milk wastes
dams, 65, 146–7, 170, and see weirs
Daphnia, 71, 76, 101
D.D.T., 58, 150–1
Dendrocoelum, 38
Deronectes, **30**
derris, 153
desmids, 11, **12**, 144
Desulphovibrio, 95
detergents, 28, 57–8, 66–7, 75, 107, 120, 134, 169
Diamesa, 32, **33**
Diatoma, 24, **25**, 105
diatoms, 11, **12**, 24, **25**, 103–4, 144, 148, 164
Dicranota, 32, **33**
Didinium, 109, **110**
dilution of effluents, 5–6, 64, 67–8, 79, 81, 95, 99, 131, 141, 157
Dinoflagellatae, 145
dragonflies, 40, **41**, 139–40, 151
Drainage of Trade Premises Act, 1937, 5
Draparnaldia, 105
drought, 15, 23, 109
dysentery, 133

E

Ecdyonuridae (flat mayflies), **29**, 32, 46, 126–7, 130, and see *Ecdyonurus*, *Rhithrogena* and *Heptagenia*
Ecdyonurus, **29**, 44, 48–9, 77, 123–6, 128–9
electroplating, 53
Elodea (canadian pond weed), 22, 24, 107
Entamoeba, 134
Enteromorpha, 140
Ephemera, **37**, 38, 46, 91
Ephemerella, **29**, 40, 44–9, 123
Ephydra, 40. **41**, 140
Epistylis, 100
Eristalis, 111, **112**, 165
erosion, see soil erosion
Erpobdella, **29**, 116, 127–8
Escherischia (*Bacterium*) *coli*, 93, 135
Esolus, 82–3, 128
estuaries, 3, 131–3
Eteone, 132
Euglena, 103, 105, 109, **110**
Eunotia, 24, **25**
Euphydatia, 38
Eurhynchium, 22, 107
Eurycercus, **39**
eutrophy,
  in lakes, 10–3, 140–4, 171
  in rivers, 24–5, 122

F

ferric hydroxide (ochre), 54, 86–7
ferrous salts, 53, 58, 62, 86, 95, 133
fertilisers, 68, 150, and see nutrient salts
fertility, see nutrient salts
Fishery Boards, 5
Fishes,
  culture of in sewage, 172–3
  feeding of, 87, 121
  as indicators of water condition, 40, 71, 163–4
  mobility of, 14–5, 39, 40, 77, 129
  in organically polluted rivers, 119–21, 128
  reactions to oxygen concentration, 119
  reactions to poisons, 77–8, 119
  selective poisoning of, 153
  spawning of, 33, 39, 98, 121, 148–9
  sudden death of, 66
  suffocation of by suspended solids, 87
  temperature tolerance of, 136–7
  tolerance to salt of, 140
fish meal, 54
fish passes, 146–7
fish species,
  barbel, 14
  bream, 11, 14, 39, 120, 141
  bullhead, 33
  carp, 136, 173
  char, 9, 11, 88, 141
  chub, 39
  dace, 39

fish species—*cont.*
eel, 40, 131
flounder, 132
goldfish, 136
grayling, 14, 16, 33, 148
gudgeon, 120
gwyniad, (*Coregonus*), 149
lake herring (*Coregonus*), 9, 11, 141
minnow, 11, 14, 33, 71, 77
omble chevalier, 11
perch, 11, 39, 120, 141
pike, 11, 39, 136, 141
rainbow trout, 71
roach, 11, 13, 39, 120, 140–1
rudd, 39
salmon, 3, 33, 40, 90, 121, 131, 136–7, 147, 150
sea-trout, 40, 87, 131
stickleback, 39, 77–8, 120, 148
stoneloach, 33
tench, 14, 120, 173
Tilapia, 173
trout, 2, 5, 11, 14, 33, 71, 74, 76, 87, 90, 121, 136–7, 141, 148
flatworms, **29**, 32, 38, 45–6, 76, 79, 83, 89, 90, 125, 127, 129–40
floods, 23, 42, 47, 50, 66, 79, 101, 106, 149
fluctuations in water level, 148–50
foam, 66–7, 134
Fontinalis, 22, 107, 125
forests, 1, 2, 146
formalin, 78, 164
Fragilaria, **12**
Frankland, Sir Edward, 58
fungi *s.s.*, 56, 99, 100, 152, 160
Fusarium, **99**, 135

G

Gammarus, 28, **30**. 40, 43–5, 72, 76–7, 89, 113, 119, 127–30, 147, 149
Gastropoda, see snails
gas wastes, 3, 6, 53–4, 57, 78, 98, 117
Gas Works Clauses Act, 1847, 3, 6
geographical distribution of animals, 14–6
Geotrichum, 100
Glaucoma, 109, **110**, 160
Glossiphonia, **29**, 116
Glossosoma, 42, 44, 128–9
Glyceria, 23, 107
Gomphonema, **25**, 81, 103–4

H

haemoglobin, 36
Halesus, 27, **39**
Haliplus, **41**, 42, 130
hardness of water, 15–6, 18, 23, 25, 28, 35, 43–4, 53, 75, 79, 99
heated effluents, 54, 64, 68, 136–9
Helmidae, **30**, 32, 40, 44–6, 83, 127–9, and see *Esolus*, *Helmis*, *Latelmis* and *Limnius*
Helmis, **30**, 47–9, 128
Helobdella, 116
Heptagenia, 47–9, 125–7
Herodotus, 3
Hildenbrandia, 24
Hippuris, (mare's tail), 22, 45
Hydra, 35, 38, **39**
Hydrobia, **39**, 132, 140
hydrochloric acid, 18, 55, 74
hydrogen sulphide (sulphuretted hydrogen), 55, 57, 62, 65, 98, 114, 133
Hydroporus, **41**, 42
Hydropsyche, **31**, 35, 43–4, 47–9, 83–4, 123, 128
Hydroptila, **39**, 44–5
hygiene, 133–5
Hypnum, 22, 83–4

I

Ice Age of Pleistocene, 16
Iceland, 173
Ilybius, **41**, 42
indicator species, value of, **162**
industrial wastes, 4, 5, 7, 53–5, 61, 81–2, 84–6, 92, 97–8, 100, 103, 125–6, 128, 139, 144, 160–1, 169–70
iron, see ferrous salts
iron bacteria, see bacteria
insecticides, 53, 58, 150–1, 153
Ischnura, 40, **41**
Isoperla, **30**, 40, 44, 47–9, 82–4, 128–9
Ithytrichia, 44

J

Joint Advisory Committee on River Pollution, 5
Juncus (rushes), 23

K

Kraft pulp mills, 92, and see wood pulping

L

lakes,
artificial enrichment of, 171–2
bottom deposits of, 146
damaged by engineering, 148–50
general limnology, 9–13, 142–4
pollution of, 140–5
prevention of evaporation from, 153–4
lakes, named,
Baikal, 9
in Central Europe, 9
Cheshire meres, 11
Esthwaite Water, 10
Fure, near Copenhagen, 144
Great Lakes of N. America, 143
Llyn Tegid (Bala Lake), 9, 149–50
Loch Lomond, 9
Lough Neagh, 9
Lowes Water, 10
in Switzerland, 11, 144
Tanganyika, 9
Thirlmere, 10
Ullswater, 88
Windermere, 9, 10, 146
of Wisconsin, 141, 152
Zürich, 144
Latelmis, **30**, 128
laundries, 54
lead, 3, 53, 74–5, 79, 80
leeches (Hirudinea), 28, **29**, 35, 38, 45–6, 79–81, 89, 90, 116–7, 127, 129, 148
Lee Conservancy, 5
Lemanea, 22, 24, 40, 79, 82–4, 105, 125, 127–8
Leptomitis, **99**
Limnaea, **31**, 32, 38, 43–4, 76, 116, 129–30, 140
Limnephilidae, 45
Limnius, 128
legal definition of pollution, 1
Leptophlebia, 40, **41**
Leptothrix, 86, 96
Leuctra, **30**, 32, 34, 40, 43–4, 46–9, 51, 80, 82–4, 123–4, 126, 128–9
lime, 55, 68
Limnodrilus, 112
Limnophora, 32, **33**
limpet, see *Ancylastrum*
Lionotus, 109, **110**
Littorina, (winkle) 132
Local Authority Acts, 5
London,
cholera in, 3
drinking water, 67
lugworm, see *Arenicola*
Lumbriculidae, 125–6
Lumbriculus, **29**
Lyngbia, 144

M

mace reed, see *Typha*
Macoma, 132
macrophytes, see weeds and moss
magnesium, 140
Manhattan, definition of, 131
mare's tail, see *Hippuris*
mayflies and nymphs (Ephemeroptera), **29**, 32–5, **37**, 38, 40, **41**, 44–9, 80, 82–4, 86–7, 90–1, 123–30, 140, 148–9, 151, 154
mercaptans, 92
mercuric chloride, 78
mesosaprobic zone, 158–9
methane (marsh gas), 55, 62, 90, 95
methods of biological sampling, 24, 26, 50–2, 164
Microcystis, **12**, 144
microhabitat, 15, 18, 43
Micronecta, 149
midges, see Chironomidae and Ceratopogonidae
milfoil, see *Myriophyllum*
milk wastes, 54, 57–8, 92, 97, 102, 118
mining and its effects, 53, 64, 78–9, 86–9, 90, 139
mites (Hydracarina), 28, **29**, 44, 46, 90, 148
molluscs, 28, **31**, 32, 35–6, **37**, **39**, 40, 43–6, 76, 79–81, 86, 89, 116, 123, 125, 127, 130, 132, 148–9
mosquito larvae, 150, and see *Culex*
moss-animalcules, see *Plumatella*
mosses, 22, 28, 40, 43–4, 83–4, 106, 125, 151
Mucor, 100
mud,
blackening of, 62, 95, 133
mixed by animals, 38, 114
Musculium, 116
mussels, see bivalve molluscs
Mya (clam), 132
Myriophyllum, 22, 23, 28, 40, 88, 139

N

Naididae, 44, 89, 113
Nais, **39**, 113
Nasturtium, (water cress), 23, 40

natural pollution, 1, 2, 9, 69
Navicula, **25**, 103, 148
Nemoura, 40, 129
Nepa, 42
Nereis, 132
Newcastle, King's College, 7
nitrates, 11, 18, 56, 58, 60, 62, 94, 103, 105, 140, 171
nitrites, 56, 58, 72
Nitrobacter, 62
Nitrosomonas, 62
Nitzschia, **25**, 81, 103–4
Notonecta, 42
Nuphar, 22, and see water lily
nutrient salts, 10, 22, 25, 68, 93, 103, 105–7, 122, 140–3, 146, 150, 170–1

O

oil, 54, 66
oilfields, 139
oligosaprobic zone, 159
oligotrophic defined, 10, 24
Oocystis, **12**
Oreodytes, **30**
Orthocladiinae, 47–9, 82–4, 115, 125, 130
Oscillatoria, **25**, 103, 141, 144
Oslo Fjord, 144–5
Ostracoda, 36, **37**
oxygen absorbed from permanganate test, 58, 60
oxygen in solution, 15–6, 20–1, 24, 27, 36, 53–5, 58–9, 65–6, 75, 86, 93–4, 97–8, 111–3, 115, 119–20, 131, 143–4, 158–9
oxygen-sag curve, 65–7, 138

P

paper industry, 54, and see wood pulping
Paramecium, 109, **110**, 111, 160, 164
parasitic worms, 133–5
pea mussels, see *Pisidium* and *Sphaerium*
Pediastrum, **12**
Penicillium, 100
Perla, 44
Perlodes, 49
pH, 18, 75, 78
Phacus, **12**, 109
phenols, 53, 56–8, 61, 72
Philopotamus, 35
Phormidium, **25**, 103
phosphates, 10, 18, 59, 60, 140-2, 170–1
photosynthesis, 21, 65–6, 105
Phragmites, 23
Physa, **39**, 40, 116
Pisidium, 28, 36, **37**
plankton,
  of lakes, 11, 12, 141, 144, 146, 171
  of rivers, 26, 35, 108
plants, see weeds
Platambus, **30**
Plectrocnemia, **31**, 33, 35, 48
Plumatella, 38, **39**
poisons, 53, 64–5, 70–85, 120, 139, 141, 145, 150–3, 156–7, 160
poliomyelitis, 3, 133
'pollution fauna', 80, **93**
Polycelis, **29**, 33, 38, 76, 82–3, **125**, 127–8
Polycentropus, 35, 44, 80, 123, **128–9**
polysaprobic zone, 158, 160–1
pond weeds, see *Potamogeton*
'ponding' of sewage filters, 56
population equivalents of effluents, 61
Potamogeton, 22, 43, 88, 106–7, 139
potassium, 10, 18, 75, 140, 170
power stations, 54, 137–8, 173
Prevention of Pollution (Scotland) Act, 1
Procladius, 115
Prodiamesa, 115
proteins, 56, 62, 97–8
Protonemura, 40, 48, 49
Protozoa, 56, 78, 93–4, 100, 108–10, 126, 133–4
Psectrotanypus, 115
Psychoda, 111, **112**, 165
Psychomyia, 44
public health, 3, 66, 135
Pygospio, 132

Q

quarrying, 53, 87

R

radioactive wastes and materials, 7, 53, 57
rain water, 18
Ranunculus (water crowfoot), 22, 23, 28, 40, 79, 88–9, 125, 128
rat-tailed maggot, see *Eristalis*
re-purification of rivers, 68
reservoirs, 149, 152–4
Rheotanytarsus, **33**, 35, 50

Rhithrogena, **29**, 32, 34, 42–4, 47–50, 83–4, 123–9
Rhizoselenia, 144
Rhyacophila, **31**, 33, 44, 48, 80, 83, 128
rivers,
  current speed of, 19
  dissolved salts in, 18–9
  ecological factors in, 18–21, 27–8
  substrata of, 15–6, 19, 22, 27–8, 42–3
River Boards, 5–8, 148, 168, 174
River Boards Act, 1948, 5
rivers and streams, named,
  Afon Hirnant, 20, 46–9
  Aire, 43
  in Ardennes, 2
  Avon, Bristol, 3, 92, 101–4, 117–8
  Avon, Hampshire, 42, 46
  Ceiriog, 46
  Churnet, 72, 80–1, 103
  Clydach, 28, 43
  Danube, 26, 166
  Dart, 2
  Ditton Brook (Liverpool), 88–9
  Don (Sheffield), 54
  Dove, 72, 80–1
  Fal, 87
  Genesee (New York), 140
  Great Driffield stream, 43–6
  Holme, 90
  in Idaho, 92
  Illinois, 118, 162
  Kalamazoo (Michigan), 118
  Lark, 98, 100, 129
  Lee, 106, 116–7, 130, 147
  in Louisiana, 78, 86
  Menominee (Michigan), 87
  Mersey, 3, 131
  in Michigan, 105
  Nidd, 43
  Nile, 26
  in Norway, 90
  in Nova Scotia, 136
  Ohio, 66, 108
  in Ontario, 15
  in Pennsylvania, 151
  Pleisse (Germany), 78
  Plym, 87
  Rheidol, 23, 80
  Rhine, 16, 67
  Ruhr, 170
  Sawdde, 43
  Scioto (Ohio), 108
  Skerne, 117
  in South Africa, 78, 111
  Spanish (U.S.A.), 90
  in Switzerland, 160
  Tamar, 132
  Tame, 101–4
  Tay, 132
  Tees, 42, 106, 117, 122–4, 126, 131–2
  Teifi, 23
  Thames, 3, 67, 131, 151
  Torry Brook, 87
  Towy, 42, 46
  Trent, 102–4, 117
  Tuolumne (California), 77
  Upper Sacramento, 73
  near Vienna, 166
  in Virginia, 151
  Vistula, 108
  Welsh Dee, 19, 46, 124–7, 132, 147, 149
  Wharfe, 43
  Ystwyth, 42, 80
River (Prevention of Pollution) Act, 1951, 6
Rivers Pollution Acts, 1876 and 1890, 4
rooted plants, see weeds,
Rotifers, 109
Royal Commissions, 3–5, 58
Royal Commission standard, 5, 157
rushes, see *Juncus*

S

Sagittaria (water soldier), 22
saline waters, see salt
Salmon Fisheries Acts, 1861 and 1865, 3
Salmon and Freshwater Fisheries Act, 1923, 5
salt (sodium chloride), 67–8, 76, 94, 97–8, 114, 132–3, 139–40
salt fly, see *Ephydra*
sampling techniques, 24, 50–2, 164
Saprobiensystem, 158–60, 165
Saprolegnia, 152
sawdust (wood fibre), 58, 90
Scandinavia, 14, 148
Scenedesmus, **12**
Schoenoplectus, (bull rush), 23
seasonal changes,
  in algal flora, 24
  in fauna, 48–50
sedges, see *Carex*
segmented worms (Oligochaeta), 28, **29**, 32, 36–7, **39**, 45–7, 79, 80, 88, 90, 125, 132, 147, 149

septic zone, 161
Sericostoma, **31**, 83, 128
sewage, 7, 54, 58, 64, 92, 123, 133–5, 141–2
  control of flies on filters, 151–2
  farms, 4
  filters, 55–6, 97–8, 109, 111
  loss of fertility in, 170–1
  treatment of, 3, 55–7, 61–3, 133
sewage fungus, 2, 78, 90, 92, 94–101, 103, 113, 121, 133, 138, 141, 155, 161, 164
shrimps (Amphipoda), 27–8, **30**, 35, 46–7, 81, 90, 125, 147–8, and see *Gammarus*
Shropshire Union Canal, 147
Sialis, (alderfly), 36, **37**, 116, 129, 149
silage, 54, 92
Silo, **31**, 32, 45
silt, 9, 19, 23, 25, 38, 90, 97, 100, 116, 124, 126, 150
silver, 75
Simocephalus, **39**
Simulium (buffalo gnat), 32, **33**, 35, 40, 44–6, 48–9, 90, 151
Sium (water parsnip), 45
slag heaps, 68
slaughter houses, 54
sludge worms, see Tubificidae
snails (Gastropoda), 27–8, **31**, 32, 36, **37**, **39**, 45–7, 72, 78, 89, 90, 113, 129, 132, 140, 147, 151–2
soil erosion, 1, 146, 150
sodium arsenite, as weed killer, 152
sodium chloride, see salt
sodium nitrite, 72
South Wales Sea Fisheries District Committee, 3
Sparganium (burr reed), 22, 23, 148
Spenser, 3
Sphaerium, **39**, 40, 116
Sphaerotilus, 56, 96–8, **99**, 100, 107, 109, 127–8, 132, 160, 172
Sphagnum, (bog moss), 23
'spills', 54, 77, 130
Spirillum, 95
Spirogyra, 105
Spirostomum, 109
sponges, 38, 90, 149
Spongilla, 38
springs, 19, 20, 86
stamping mill grit, 87
standards for effluents, 4–6, 157, 168
starwort, see *Callitriche*
Staurastrum, **12**, 144
stearyl alcohol, 153
steel works, 54
Stenophylax, **31**, 47, 129
Stentor, 109
Stephanodiscus, **12**
Stigeoclonium, **25**, 81, 103–4, 140, 164
stoneflies (Plecoptera), **30**, 32–5, 38, 44–9, 76, 80, 83–4, 90–1, 113, 123, 125–30, 137, 140, 148–9, 151, 175
storm overflows, 66
streams, named, see rivers
Streptococcus, 135
Stylaria, **39**
sugar factories, 54, 97, 100, 129
sulphates, 55–6, 58, 62, 66, 76, 86, 95, 98, 133
sulphides, 53, 57–8, 62, 64–6, 75, 92, 95, 98, 119, 133, 144, 158–9
sulphites, 53–4, 56, 68, 86
sulphur, see sulphides and sulphates
sulphur bacteria, see bacteria
sulphuretted hydrogen, see hydrogen sulphide
sulphuric acid, 55, 78, 98
surface run off, 1, 19, 146
Surirella, **25**, 103
suspended solids, 56, 64–6, 86–94, 107, 113–6, 121
swamps, 87
Synedra, **25**, 144
Synura, **12**, 144

T

Tabellaria, **12**, 24, 144
Taenia, 134
Taeniopteryx, 40
tadpoles, 76
tanning, 53–4
Tanypodinae, **33**, 48–9, 83–4, 115, 130
Tanytarsus, **37**, 38, 116, 124, 130, and see *Rheotanytarsus*
tar and tar acids, 53, 74, 78, and see phenols
temperature, 15, 19–20, 27, 68, 136–8
  effect on oxygen, 21, 66, 119
  effect on toxicity, 75, 139
Tetraëdron, **12**
Tetraspora, 105
textile wastes, 54, 57, 68, 90
Thames Conservancy, 5
Thiobacillus, 57, 98
thiocyanates and thiosulphates, 54, 57, 74, 98

tidal waters, 3–4, 131–3
Times, The, 3
tin mining, 3
Tipulidae (craneflies), 32, 45
toxicity, see poisons
Trachelomonas, 111
treatment of effluents, 3–5, 53–4, 63, 84, 128, 133–4, 168–74
Trichiurus, 134
Trichoptera, see caddis
Trichostrongylus, 134
Trocheta, 116
Tubifex, **37**, 112
Tubificidae (sludge worms), 2, 36, **37**, 38, 72, 76, 81, 89, 90, 93–4, 112–8, 121, 129, 137, 140, 165
turbidity, 87, 141, 143–4
Typha, (mace reed), 23
typhoid fever, 133

U

Ulothrix, **25**, 103
Ulvella, **25**, 103–4, 122
underground waters, 86
Unio, 36

V

Vallisneria, 137
Valvata, 36, **37**
Vaucheria, 22
virus diseases, 133
Vorticella, 109, **110**, 111, 164
Vorticellidae, 56, and see *Carchesium*, *Epistylis* and *Vorticella*

W

Walton, Isaak, 2
water blooms, see algae
water boatmen, **41**, 42, 76, 140, 149
water celery, see *Berula*
water cress, see *Nasturtium*
water crowfoot, see *Ranunculus*
water lily, 22, 107
water louse, see *Asellus*
water parsnip, see *Sium* and *Apium*
Water Pollution Research Laboratory, 6, 169
water scorpion, see *Nepa*
water slater, see *Asellus*
water soldier, see *Sagittaria*
water works, 11, 141, 144, 152–3, 171
weeds (rooted plants), 10, 22
  communities of, 22–3, 137
  control of, 152
  damaged by animals, 27
  factors affecting, 22
  as habitat for animals, 38–43
  instability of, 23
  in polluted water, 79–80, 106–8, 141
weirs, 66, 147, 170
West Riding Rivers Board, 90
winkle, see *Littorina*
wood fibre, see sawdust
wood pulping, 86, 92, 98–9, and see paper industry and sawdust
worms, see segmented worms

Z

zinc, 3, 53, 76, 79, 80
zones of organic pollution, 93, 158–60
Zoogloea, 56, 97–8, **99**, 166
Zürich, 166, and see lakes

AUTHOR INDEX

Where the page number is shown in italics the author is included in the epithet *et al.*

Alabaster, J. S., 119–*20*, 139
Alexander, W. B., 131
Allan, I. R. H., 120
Allen, L. A., 54, 74, 134–5
American Public Hlth. Assoc., 57
Anderson, B. G., 76

Bachmann, H., 76
Bahr, H., 96, 98, 135
Ball, J. N., 90
Bartsch, A. E., 93, 116, *141*
Bassindale, R., 75, *131*
Beaujean, P., *166*
Berg, K., 15, 98, 114, 116, 142, 144
Birrer, A., *76*
Bishop, L. M., *141*
Blezard, N., *54*, *74*, *134*
Blum, J. L., 82, 101, 103, 105
Boyd, W. L., *141*
Braarud, T., 144
Brinley, F. J., 103, 108, 160
Brown, F. M., *90*
Burgess, S. G., 58
Butcher, R. W., 14, 19, 21–2, 24, 26, 42, *46*, 66, 72, 80, 88, *92*, 96–106, 116–9, 122–3, 129, 137, 163, 165–6

Campbell, M. S. A., 92, 103, 116, 160
Carpenter, K. E., 14, 70, 75–6, 79
Claassen, P. W., 140
Clemens, H. P., 140
Cole, A. E., 71, 74, 76, 87
Collett, R., *150–1*
Collman, R. V., 5
Cooke, W. B., 56, 96, 100

Davies, H. S., 50
Degens, Ir. P. N., 67, 107
Dept. of Scientific and Industrial Research, 131
Domogalla, B., 152
Doudoroff, P., 71, 75–7
Downing, A. L., *67*
Downing, K. M., 75, 139
Drooz, A. T., 151
Dunlop, S. G., 133–4

Eastham, L. E. S., 38
Edwards, R. W., 38, 103, 105, 114
Elkins, G. H. J., *75*, *120*
Ellis, M. M., 71, 76, 162
Elson, P. F., 150
Everhart, W. H., 150

Fair, G. M., *141*, *144*, *152–3*, *160*
Fitch, C. F., 141
Fjerdingstad, E., 101, 103, 105, *142*, *144*, 159
Flaigg, N. G., 142

Gameson, A. L. H., 67, 170
Gardiner, A. C., *10*, 74, *144*, *153*
Gardner, J. A., 136
Garner, J. H., 90
Gaufin, A. R., 92, 112, 116, 159, 161–2, 165
Gorham, E., 18
Gortner, R. A., *141*
Gray, E. A., 109
Greenbank, J., 153
Greenshields, F., *10*, *144*, *153*
Griffiths, R. B., 133–4
Grindley, J., *42*, *46*, *57*, *92*, *116–9*, *150–1*
Gussew, A. G., 99, 101

Haempel, O., 76
Harmic, J. L., 163
Harrison, A. D., 78, 111
Harvey, J. V., 100
Hasler, A. D., 141–2, 144
Hassler, W. W., 150
Hawkes, H. A., 95, 133
Hemens, J., *75*, *119–20*, *139*
Hentschel, E., 159
Herbert, D. W. M., 73, 75, *119*, 120, *139*, 140
Herman, E. F., *141*
Heuts, M. J., 136

Hey, D., 173
Hoffmann, C. H., 150–1
Huet, M., 2, 14, 71, 165–6
Huntsman, A. G., 136
Hynes, H. B. N., 15, 35, 126, 175

Illies, J., 14, 16, 26, 35
Ilzhöffer, 172
Imhoff, K., 170–1
Ingram, W. M., 141
Isaac, P. C. G., 7, 172–3

Jaag, O., 68, 150
Jenkins, S. H., 55
Jónasson, P. M., 51
Jones, J. R. E., 23, 28, 35, 42–3, 46, 71–2, 74–7, 79–80, 88, 119, 139
Jones, J. W., 90
Jones, W. H., 140

Kaiser, E. W., 112
Kamphius, A. H., *67, 107*
Kaplovsky, A. J., 163
Katz, M., 71, 75–7, *171*
Kerswill, C. J., 150
Kisskalt, K., 172
Klein, L., 53, 55, 58, 60, 156, 168
Klingler, K., 72
Kolkwitz, R., 101, 158–60, 165
Kommer, J. D., *67, 107*

Lackey, J. B., 96–7, 108–9, 111, 141, 152, 171
Lafleur, R. A., 79, 86, 92, 139
Larsen, K., 87
Laurent, P. J., 76
Laurie, R. D., 79, 80
Lea, W. L., 60, 171
Leclerc, E., *166*
Leonard, J. W., 153
Liebmann, H., 2, 95–8, 100–1, 103, 105, 108–9, 111, 133, 140, 158–60, 164
Liepolt, R., 111, 133, 166
Lindroth, A., 119
Lloyd, L. L., 55–6, 111
Lodge, M., 172–3
Longwell, J., *42, 105, 116–7, 119, 122–3*
Lovett, M., 3, 58, 60, *90*
Lowndes, A. G., 78
Ludwig, H. F., 172

Macan, T. T., 13, 19, 52, 142, 172
Mack, B., 101, 103
Mackenthun, K. M., 141, 152
Mallman, W. L., 135
Mann, H. T., *75, 120,* 140
Mann, K. H., 116, 137
Margalef, R., 24
Marsson, M., 158–60, 165
Merkens, J. C., 73, 75, 139
Meuvis, A. L., 136
Ministry of Health, 6
Ministry of Housing, 58, 107, 171
Mohr, J. L., 108
Moon, H. P., 35, 149
Mortimer, C. H., *172*
Mossewitsch, N. A., 99, 101
Mottley, C., McC., 52
Moyle, J. B., 152
Mueller, P. K., 78
Muirden, M. J., 150–1
Müller, K., 148

Naylor, E., 137
Newton, L., 76, 79–80
Nicholson, H. P., 150
Nitardy, E., 166
Nowak, W., 159, 161, 165

Ohle, W., 142, 171
Olsen, S., 87
Olszewski, W., 95, 155
Oswald, W. J., 172
Owen, R., 171

Painter, H. A., 56
Pasley, S. M., *135*
Patrick, R., 103, 158, 162
Paul, R. M., 57, 59, 74, 77, 79, 87, 90, 150
Pearsall, W. H., 10, 13, 144, 146, 153, 171
Pehrson, S. O., 98
Pennington, W., 146
Pentelow, F. T. K., 6–8, *21,* 42, 46 64, 72, *75,* 80, 87, 92, *96–8, 100, 105,* 116–9, 123, 129, 131
Percival, E., 15, 23, 35, 42–4, 51
Phelps, E. B., 58–61, 65–6, 131, 133 155, 157, 168
Pielou, D. P., 150
Pirie, N. W., 173
Postgate, J., 57, 95, 98
Prescott, G. W., 141, 144, 152
Pringsheim, E. G., 96
Pruthi, H. S., 65

Rainwater, J. H., *52*
Rasmussen, C. J., 54, 92, 119, 121
Rayner, H. J., *52*
Redeke, H. C., 71, 77, 90, 92
Rees, W. J., 137
Reese, M. J., 79, 80
Reid, G. W., 142
Rennerfelt, J. G. V., 60
Richardson, R. E., 113, 116, 118, 162, 165
Ricker, W. E., 15
Roberts, C. H., 57
Rogers, C. F., *141*
Rohlich, C. A., *141*
Rudolfs, W., 56–7

Sawyer, C. N., 141
Schmitz, W., 14, 140
Schneller, M. V., 1
Silverman, P. H., 133–4
Simmonds, I. G., 90
Sládeček, V., 56, 109
Southgate, B. A., 6, 7, 53, 58, 67, 71, 74–5, 93, 95, *131*, 133–4, 136, 138, 169
Spitta, O., 95, 155
Šrámek-Hušek, R., 108–9, 159–60, 164
Stammer, H. A., 76
Steeman Nielsen, E., 172
Steinmann, P., 161
Stjerna-Pooth, I., 99, 100
Stopford, S. C., 132
Stuart, T. A., 90
Stundl, K., 92, 101
Suckling, E. V., 57, 95, 155
Surbeck, G., 161
Surber, E. W., 87, 118, 150–1, 158

Tarzwell, C. M., 92, 112, 116, 159, 161–2, 165
Tate Regan, C., 88
Thienemann, A., 16, 114–5, 124, 175
Thomas, E. A., 109, 141, 144, 160, 166, 170–1
Tilden, J. E., *141*
Timmermans, J. A., *166*
Tomlinson, T. G., 150–1
Truesdale, G. A., *67*
Turaboyski, L., 108
Turing, H. D., 7, 71

United States Dept. of the Interior, 154

Vallin, S., 98–100, 115
Van Der Zee, H., *67*, *107*
Van Horn, 92
Van Oosten, J., 87, 143

Wallen, I. E., 87
Walshe, B. M., 35–6
Wang, W. L. L., 133–4
Wattie, E., 96–7
Weber, H., *76*
Welch, P. S., 9, 89, 142
Wesenberg-Lund, C., 35–6
Westlake, D. F., 38, 103, 105, 114
Wheatland, A. B., *54*, *74*, *134*
Whipple, G. C., & M. C., 141, 144, 152–3, 160
Whitehead, H., 15, 23, 35, 42–6, 51
Wiebe, A. H., 78, 92, 108, 116–7
Williams, E. H., *57*
Wilson, J. N., 92, 98, 109, 151
Wilson, W. L., 53
Wisdom, A. S., 1, 5, 6
Woker, H., 70, 73, 75–6
Woodley, J. W. A., *21*, *96–8*, *100*, *116*, *129*
Worthington, E. B., 13, 19, 142, *172*
Wuhrmann, K., 70, 73, 75–6
Wurtz, C. B., 163

Young, L. A., 150

Zehender, F., *75*